SpringerBriefs in Earth System Sciences

South America and the Southern Hemisphere

For further volumes:
http://www.springer.com/series/10032

Pedro José Depetris · Andrea Inés Pasquini
Karina Leticia Lecomte

Weathering and the Riverine Denudation of Continents

Springer

Pedro José Depetris
Centro de Investigaciones en Ciencias de la Tierra-CICTERRA (CONICET-Universidad Nacional de Córdoba)
Córdoba
Argentina

Andrea Inés Pasquini
Centro de Investigaciones en Ciencias de la Tierra-CICTERRA (CONICET-Universidad Nacional de Córdoba)
Córdoba
Argentina

Karina Leticia Lecomte
Centro de Investigaciones en Ciencias de la Tierra-CICTERRA (CONICET-Universidad Nacional de Córdoba)
Córdoba
Argentina

ISSN 2191-589X ISSN 2191-5903 (electronic)
ISBN 978-94-007-7716-3 ISBN 978-94-007-7717-0 (eBook)
DOI 10.1007/978-94-007-7717-0
Springer Dordrecht Heidelberg New York London

Library of Congress Control Number: 2013950358

Printed on acid-free paper

Springer is part of Springer Science+Business Media (www.springer.com)

Preface

Continental denudation has been the center of attention of a reduced and select group of outstanding specialists for several decades. The foci have been placed in queries such as how much solid and dissolved material is annually removed from the continents by rivers and how much water is involved in the process?; which are the forcing variables and processes in continental wearing away?; is climate more important than lithology in defining the rate of denudation?, where and when?; how much height do continents recuperate through isostatic rebound as a consequence of denudation?; are weathering and denudation connected in a straightforward way? These and other similar questions became particularly important several years ago, when it turned out that weathering and denudation were linked to climate change, nowadays or in the geological past. The joint action of weathering and denudation has played, for example, a significant role in the formulation of the amazing hypothesis of the "Snowball Earth," which appears to have occurred occasionally about 650 million years ago. On the other hand, there exists the additional anthropogenic impact that adds complexity to the current scenario because it may amplify erosion in ill-managed areas and, hence, increase sediment retention behind dams.

We have written this Brief Monograph, using classic references and case studies, trying to direct the reader's attention to the simple fact that continental denudation begins with weathering. Rivers constitute the amazingly efficient conveyor belt system that performs the main task in transferring most of the solid debris and dissolved substances thus produced, to the coastal ocean. At the same time, we must not disregard the fact that, in so doing, they contribute to sustain a very dynamic and diverse life-supporting system, in the rivers themselves as well as in estuaries and the adjacent marine environment.

On account of its conciseness, this brief book should be viewed as a stepping stone for those who wish to pursue this subject further on and, also, as an invitation to some deeper digging into the weathering-denudation system in a multidisciplinary way.

Acknowledgments

The authors wish to acknowledge the continued support of Argentina's Consejo Nacional de Investigaciones Científicas y Técnicas (CONICET) and of the Universidad Nacional de Córdoba (Córdoba, Argentina). Both institutions sustain the Centro de Investigaciones en Ciencias de la Tierra (CICTERRA, http://www.cicterra-conicet.gov.ar/), where this monograph was conceived and created. The gratitude is also extended to Jorge Rabassa, current Director of CONICET's CADIC in Ushuaia (Argentina) and enthusiastic promoter of this series, who extended a warm invitation and the necessary encouragement to produce this manuscript. Thanks are also due to Marcela Cioccale, Marcos Chaparro and Gabriela Zanor, who generously put at our disposal abundant photographs in order to select graphic material adequate to illustrate processes described in the text.

Several editors have kindly granted the necessary permission to reproduce graphic material and thanks are extended to all. The authors are specially obliged to Australia's CSIRO (http://www.publish.csiro.au/pid/5955.htm) and to the UK's Cambridge University Press (http://www.cambridge.org/ar/academic/subjects/earth-and-environmental-science/geomorphology-and-physical-geography/river-discharge-coastal-ocean-global-synthesis?format=PB).

Contents

Chapter 1
Introduction

Abstract Denudation, volcanism, and tectonics are intertwined Earth system processes that constitute the main driving forces intervening in shaping the Earth's landscape. Clearly, the wearing away of the Earth's surface cannot occur unless a series of synergistic processes, collectively known as "weathering," are initiated. This term, in use for a long time, promotes the idea that climate (weather) always plays a major role in rock breakdown; since this is not the case in every instance the change for "rock decay" has been proposed recently. At any rate, the linkage between weathering and denudation is not straightforward because the latter may be limited by the former ("weathering-limited denudation") or, in contrast, it may be restricted by the hindered transport of the weathering-produced debris ("transport-limited denudation"). In addition to these possible scenarios, two new approaches have been gaining growth in the recent past: one is the study of the "regolith" as a convenient research framework, and the other is the notion of "the critical zone," where the dynamic interaction with the atmosphere and vegetation is emphasized and added to the materials collectively defined as regolith.

Keywords Denudation • Climate • Rock decay • Regolith • Saprolite • Colluvium • Alluvium • Aeolian deposits • Weathering-limited • Transport-limited • Critical zone

1.1 On Weathering and Denudation

The Earth's surface is perceived as dynamic because the remarkable effects of denudation, coupled with tectonics, are constantly changing its landscape. Weathering, erosion, and mass wasting must act before one can appreciate the effects of denudation; sometimes, they are spectacular and sometimes they are subtle, almost imperceptible. At any rate, the weathering of rocks is the preliminary process which modifies and destabilizes the uppermost layers of the Earth's land surface.

P. J. Depetris et al., *Weathering and the Riverine Denudation of Continents*,
SpringerBriefs in Earth System Sciences, DOI: 10.1007/978-94-007-7717-0_1,

Mineral and rock weathering is clearly the term that has been used for a long time, which describes the initiation of continental wear away. It refers to a series of processes that collectively interact and contribute in variable proportions to shape up the Earth's landforms and landscapes and, hence it has been a central concept in geomorphology.

We use in this monograph the traditional approach that divides the notion of mineral/rock decay into mechanical (or physical), biological, and chemical processes, but we tackle the problem in this way because we still lack a more evolved approach that collates all the processes that participate in the alteration of the exhumed Earth's crust. Dictionaries describe the term "*weathering*" as the action of the weather on any kind of material which is exposed to it. In geology, the term is used specifically to describe the breaking down of minerals and rocks through direct contact with the Earth's atmosphere, and waters. In the geological sense, however, the "long-term action" of weather is implied, as it happens in many definitions used in the Earth Sciences. At any rate, the ultimate action of weathering is to free materials (Fig. 1.1).

The difficulties involved in what is understood by *weathering* are currently being considered, and new ways are being explored. For example, the term itself has had a commanding influence on the evolution of research, particularly when one considers the implication that the weathering of any rock type will follow closely the path set by climate. It is more likely that the weathering of a particular rock, with a relatively constant composition, will be invariable in nature but variable in rate, given the changeability of climate forcing (Hall et al. 2012). To

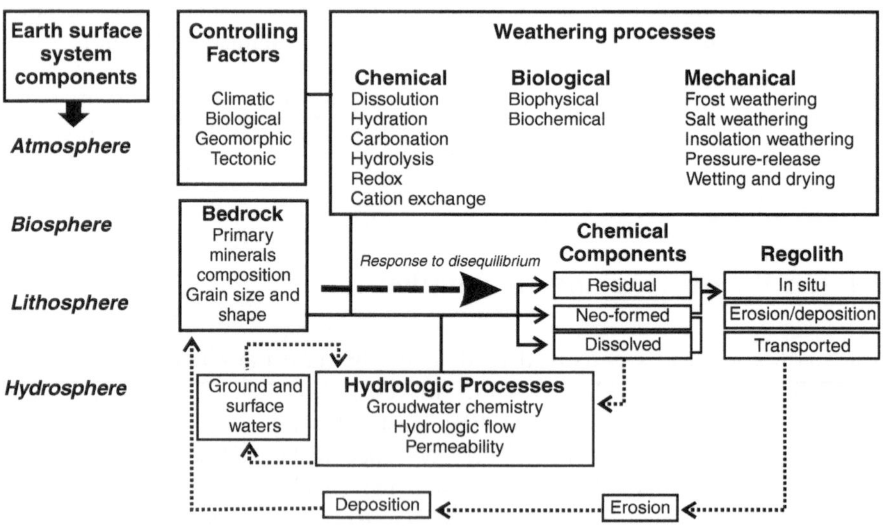

Fig. 1.1 The foremost controls on weathering and the chemical composition of the regolith. Modified from McQueen (2009). © CSIRO 2008. Published by CSIRO publishing, Collingwood, Victoria, Australia, http://www.publish.csiro.au/pid/5955.htm. Reproduced with permission

bypass the constrain set by the term, Hall et al. (2012) have proposed the use of "rock decay," which is a denomination that, in their view, reflects the reality of what happens on the Earth's surface more accurately. At any rate, our understanding on the subject has been deepened and broadened during the last few decades, thanks to the input of geochemistry, biochemistry, and geobiology. Also significance has been the growth of laboratory studies, which has brought increased understanding on the conditions that lead to physical or mechanical rock breakdown. Today, as we try to show in the following pages, increasing knowledge of, and interest in fields such as the studies of the performance of stones used as building materials, and the conservation of ancient buildings, are also important foundations for the currently available information.

The layer of *in situ* and transported rock debris that mantles landscapes across the land is known as *regolith* and is a ubiquitous feature at the Earth's surface that is laterally extensive and can reach significant thickness in some places. This regolith, that "includes fractured and weathered basement rocks, *saprolites*, soils, organic accumulations, glacial deposits, *colluvium*, *alluvium*, *evaporitic sediments*, *aeolian deposits* and ground water" (Scott and Pain 2008a) is the loose material available for transportation that constitutes the debris removed from the continents and delivered to coastal seas and oceans. The study of regolith is receiving growing multidisciplinary attention in as much as it has many linkages with aspects that are important for the sustainability of life on Earth (Scott and Pain 2008b). Moreover, if the outer extent of vegetation is added to the regolith layer, we define a wider area that is currently known as the *critical zone* (CZ). Clearly, life is sustained by this delicate skin of the planet, which is receiving mounting and well-deserved attention (Brantley et al. 2007).

The notion of weathering and transport-limited systems was first launched into scientific literature by Carson and Kirkby (1972). These ideas have proved to be essential for the understanding of weathering as a synergistic and complex process. In weathering-limited circumstances, there is an efficient transport of weathered products (particulate and dissolved phases) away from the surface, thus preventing the development of any more than thin and irregularly distributed soils. Typically, such weathering-limited scenarios are dominated by biologically mediated mechanical (or physical) weathering processes.

Conversely, transport-limited situations are dominated by the buildup of rock debris and a soil cover over the exposed rock surface, as there is only a sluggish or discontinuous removal of the weathering products. The deposition on the bare surface of allochthonous debris (e.g., aeolian, glacial, etc.) is another way of generating transport-limited conditions. Viles (2013) has explored the synergistic nature of weathering and, in the process of analyzing the linkage between weathering and erosion, quoted the model of Gabet and Mudd (2009), which suggests that chemical weathering rates increase markedly with increasing erosion, up to about 100 t km^{-2} yr^{-1}. At higher denudation rates, chemical weathering exhibits less of an increase. Adding further to this nonlinear image, their model also predicted that the greatest amount of chemical weathering should be expected in drainage basins that have an average soil thickness of ~50 cm.

As it arises from the previous paragraphs, denudation is understood in the Earth Sciences as the long-term synergism of various processes that produce the wearing away of the Earth's surface leading to a sustained decrease in elevation, thus altering the relief of landforms. Endogenic processes such as volcanoes and crust exhumation through continental uplift expose the Earth's relatively thin outer layer to exogenesis, which—as pointed above—involve the weathering of minerals and rocks, the erosion of exposed materials, and mass wasting. Therefore, in order to free mineral grains originally firmly bound in hard rocks or more loosely in friable sediments, so as to make them amenable for transportation by water, wind, or ice, denudation makes use of mechanical, biological, and chemical processes that participate in the removal from continental landmasses of both, solid particles and dissolved phases. Particles are transported from its original source and then they are stored in ocean basins or in coastal zones for geologic times until plate tectonics determine the subsequent stage in the cycle (e.g., exhumation as sedimentary or metamorphic rocks, eruption as melted magma). Dissolved materials, on the other hand, end up in seas and oceans, contributing to their intricate chemistry and sustaining a complex biological network.

Current information attributes to the world's rivers the delivery of 36,000 km^3 of water (Milliman and Farnsworth 2011); at this rate, one volume of ocean water becomes recycled in roughly 40,000 years. On a worldwide basis, many authors have probed into the role of rivers as effective transport agents to the world ocean; classic contributions are those of Clarke and Washington (1924), Livingstone (1963), Holeman (1968), Garrels and Mackenzie (1971), Baumgartner and Reichel (1975), Martin et al. (1980), Meybeck (1982). More recently, Milliman and Meade (1983), Degens et al. (1991), Milliman and Syvitski (1992), Probst (1992), Walling and Webb (1996), and Ludwig and Probst (1998), contributed additional information to this fascinating subject. Recently, Milliman and Farnsworth (2011) have reported on the discharge of rivers to world oceans after collecting for many years a very extensive (and impressive) dataset.

In this monograph, we seek to examine the processes that participate in the riverine denudation of continents; we do not attempt to collate all relevant data on rivers that discharge directly into the global ocean; such task has been recently and worthily tackled by Milliman and Farnsworth (2011) and we have abundantly used such reference throughout this book. Our aim is, rather, to review current knowledge on the processes of weathering and denudation that participate in continental wearing down. The aim is, in other words, to convey the image that such processes are links in a chain with many and far-reaching implications, such as the change of the Earth's climate, nowadays or in the distant geological past (Lenton et al. 2012).

Glossary

Aeolian deposit: Fine sediments transported and deposited by wind. Loess, dunes, desert sand, and some volcanic ash are included in this category.

Alluvium: Material deposited by rivers, usually forming floodplains and deltas, which is typically most extensively developed in the lower part of the course of a river; consists of silt, sand, clay, and gravel and often with a significant proportion of organic matter.

Colluvium: Soil and debris that accumulate at the base of a slope by mass wasting or sheet erosion. It usually includes unsorted angular fragments, and may contain slabs of bedrock that indicate both their place of origin and that slumping was the process of transportation. At the edges of valleys, colluvium may be mixed with and almost indistinguishable from alluvium.

Critical zone: It is the zone critical for the sustenance of life on Earth, which encompasses the lowermost groundwater to the atmosphere that meets the Earth; it is the zone where the lithosphere intersects with the biosphere, hydrosphere, and atmosphere.

Evaporitic sediment: Sediment soluble in water that results from concentration and crystallization by evaporation from an aqueous solution. Marine and non-marine are the types of evaporate deposits. The latter are found in standing bodies of water such as lakes (playas).

Regolith: Layer of loose, heterogeneous material covering solid rock. It includes dust, soil, broken rock, and other related materials and is present on Earth, the Moon, Mars, some asteroids, and other terrestrial planets and moons.

Saprolite: Is a chemically weathered bedrock surface, found in the lower zones of soil profiles. Deeply weathered profiles are widespread on continental landmasses, between 35° N and 35° S.

References

Baumgartner A, Reichel E (1975) The world water balance. Mean annual global, continental and maritime precipitation, evaporation and run-off. Elsevier, Amsterdam

Brantley SL, Goldhaber MB, Ragnarsdottir KV (2007) Crossing disciplines and scales to understand the critical zone. Elements 3(5):307–314

Carson MA, Kirkby MJ (1972) Hillslope form and process. Cambridge University Press, Cambridge

Clarke FW, Washington H (1924) The composition of the earth's crust. US Geol Surv Prof Pap 127:117, Washington, DC

Degens ET, Kempe S, Richey JE (1991) Biogeochemistry of major world rivers. John Wiley and Sons, Chichester

Gabet EJ, Mudd SM (2009) A theoretical model coupling chemical weathering rates with denudation rates. Geology 37:151–154

Garrels RM, Mackenzie FT (1971) Evolution of sedimentary rocks. WW Norton and Co, New York

Hall K, Thorn C, Sumner P (2012) On the persistence of “weathering”. Geomorphology 149–150:1–10

Holeman JN (1968) Sediment yield of major rivers of the world. Water Resour Res 4:737–747

Lenton TM, Crouch M, Johnson M et al (2012) First plants cooled the ordovician. Nat Geosci 5:86–89

Livingstone DA (1963) Chemical composition of rivers and lakes.US Geol Surv Prof Pap 440-G:1–64, Washington, DC

Ludwig W, Probst JL (1998) River sediment discharge to the oceans: present-day controls and global budgets. Am J Sci 298:265–295

McQueen KG (2009) Regolith geochemistry. In: Scott KM, Pain CF (eds) Regolith science. Springer, Dordrecht

Martin JM, Burton D, Eisma D (1980) River inputs to ocean systems. UNEP-UNESCO, Geneva

Meybeck M (1982) Carbon, nitrogen and phosphorous transport by mayor world rivers. Am J Sci 282:401–501

Milliman JD, Farnsworth KL (2011) River discharge to the coastal ocean: a global synthesis. Cambridge University Press, Cambridge

Milliman JD, Meade RH (1983) World-wide delivery of river sediments to the ocean. J Geol 91:1–21

Milliman JD, Syvitski JPM (1992) Geomorphic/tectonic control of sediment discharged to the ocean: the importance of small mountainous rivers. J Geol 100:525–544

Probst JL (1992) Geochimie et hydrologie de l’erosion continental. Mechasnisms, bilan global actuel et fluctuations au cours des 500 millonsd’annes. Sci Géol Bull 94:161

Scott KM, Pain CF (2008a) Introduction. In: Scott KM, Pain CF (eds) Regolith science. CSIRO Publishing and Springer, Collingwood

Scott KM, Pain CF (2008b) Regolith science. CSIRO Publishing and Springer, Collingwood

Viles HA (2013) Synergistic weathering processes. In: Shroder JF (ed) Treatise on geomorphology, vol 4. Academic Press, San Diego

Walling DE, Webb, BW (1996) Erosion and sediment yield: a global overview. IAHS Publ 236:3–19

Chapter 2
The Commencement of Continental Denudation: Mechanical Weathering

Abstract Although weathering processes are increasingly perceived as of a synergistic nature, mechanical (or physical) weathering can be viewed as the most frequent commencement of continental denudation. The action of temperature, humidity, and pressure interact to breakdown minerals and rocks, thus increasing the specific surface and facilitating the action (subsequent or concurrent) of biological and chemical agents. In order to understand the processes involved in mechanical weathering, field observations are necessary but not sufficient to fully appreciate the mechanisms of rock mechanical breakdown. In the past decade, laboratory experimentation has provided a wealth of information on the physics involved to reduce the size of outcropping rock blocks, so as to finally allow the transportation of continental rock debris to its deposition on the ocean floor, which is ultimately the final stage of the Earth's exogenous cycle.

Keywords Physical weathering · Frost weathering · Salt weathering · Insolation weathering · Pressure-release · Wetting and drying · In situ regolith · Crystallization pressure · Thermal expansion · Exfoliation

2.1 Introduction

Mechanical or physical weathering involves the breakdown of rocks and minerals through direct contact with atmospheric conditions, brought about by a variety of causes such as heat, water, ice, and pressure. An important initial remark on the characteristics of weathering is that it occurs in situ, or "with no movement." Erosion is the exogenous process that involves the movement of rock debris and minerals by agents such as water, ice, snow, wind, and gravity and thus it should not be confused with weathering. *Regolith*, on the other hand, is a term that involves the weathering of rocks and minerals but it may (or may not) include the erosion, transport, and/or deposition of the older rock debris. In contrast, an in situ regolith consists of physically broken mineral material that generally has endured

P. J. Depetris et al., *Weathering and the Riverine Denudation of Continents*, SpringerBriefs in Earth System Sciences, DOI: 10.1007/978-94-007-7717-0_2,

some incipient degree of chemical alteration, and whose mineralogy and chemistry is close to the original rock substrate (Scott and Pain 2009).

In physical weathering, some forces originate within the rock or mineral, while others are applied externally. Both of these stresses ultimately lead to strain and the rupture of the rock. Although it implicates rock or mineral disintegration without the occurrence of any significant chemical alteration, incipient chemical change may, however, diminish the strength of a rock to a level at which the stresses of mechanical weathering are sufficient to promote rock breakdown. The sequence that these processes may follow or the extent to which each one plays a role is a complex subject that is being investigated. For example, rock fractures caused by physical weathering will increase the surface area exposed to chemical action and, in turn, chemical action in fissures can aid the disintegration process. Moreover, rock and mineral breakdown are often promoted by living organisms whose zone of action is widely known as the *biosphere*. The involvement of organisms in material breakdown is so distinctive that their effect on the *lithosphere* determines a separate category in weathering, with biochemical and biophysical processes as usual subcategories (See Chap. 3). From the physical point of view, for example, lichens or fungi are considered significant agents of biological weathering as they are deemed capable of disaggregating minerals or rocks and of producing substances that induce a chemical attack.

It must be added here that since it is possible to date with reasonable accuracy the exposure time of building stones in order to assess their weathering rate, such investigations have supplied a wealth of information on the response of different rocks to physical as well as chemical agents (e.g., Camuffo 1995; Papida et al. 2000).

In this chapter, the main processes that involve the mechanical breakdown of minerals and rocks are succinctly reviewed, using some examples supplied by scientific literature to assist in the assessment of the state-of-knowledge in this particular field.

2.2 Frost Weathering

Frost weathering is a short title for a group of processes that are also known as water-based, low-temperature weathering: freeze-thaw weathering, hydration shattering, ice crystal growth, and hydraulic pressure (Bland and Rolls 1998). Recent experimental studies have increased significantly the knowledge on the physical characteristics of these processes (Weber et al. 2012).

Freeze-thaw weathering occurs by the freezing of in-place water or by water migration and ice growth (Fig. 2.1). In the first case and simply expressed, freeze-thaw begins when water enters fissures in bare rocks; during freezing nights, the frozen water expands against the rock. Ice melts when temperatures rise and this causes more room for additional water to enter the crack. Over time, the expansion

Fig. 2.1 Example of frost weathering in the Famatina range (La Rioja, Argentina, ~29 ° S and ~67 ° 42' W). Rock fracturing in the "Negro Peinado" metamorphic formation probably occurred due to the combined action of freeze-thaw and ice crystal growth. Photograph by M. A. Cioccale

and contraction of water breaks the rock. Frozen water occupies ~9 % more volume than its equivalent in liquid water. Under the best possible conditions, at −22 °C, a theoretical pressure of 207 MPa is reached, which is much greater than rock tensile strength (~10 MPa). In the field, however, this pressure would seldom be reached for several reasons that we will not describe here. Nonetheless, it seems that high pressure triggered by water freezing could be maintained under certain circumstances (Bland and Rolls 1998):

- A freezing rate of more than −0.1 °C min^{-1} in a water-filled crack (e.g., 1 mm wide and 10 cm deep), from the top downwards, would allow ice to act as a seal. In such situation, pressures greater than 19.6 MPa would develop, for example, in granite (i.e., a saturated nonporous rock).
- "Frost bursting" can occur through the very rapid conversion of supercooled water to ice (~−6 °C) because the rapid freezing rate would prevent loss of pressure toward the water entry point.

  In contrast with the previous mechanism, freeze-thaw weathering occurs by water migration and ice growth involves flowing water that reaches a freezing zone. Laboratory and field studies link:

  - Rock properties, such as pore size (i.e., affects freezing temperature); fracture geometry and toughness; and permeability (i.e., controls water flow at subzero temperatures).
  - Temperature regime, which stimulates fracture propagation at temperature ranges of −5 to −15 °C and, also, when cooling rates are less than −0.1 to −0.5 °C h^{-1}.
  - Moisture is an effective factor when there is abundant water supply.

Freeze-thaw weathering is frequently mentioned as a significant agent of landform development in high latitudes and in mountainous regions (Fig. 2.2). This mechanism was not widely accepted by all specialists; Hall (1995), for example, argued that in view of the scarcity of data regarding key factors, such as rock

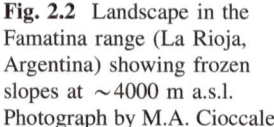

Fig. 2.2 Landscape in the Famatina range (La Rioja, Argentina) showing frozen slopes at ~4000 m a.s.l. Photograph by M.A. Cioccale

temperature and interstitial rock moisture content data, the assertion was largely qualitative. He was particularly critical of the claim that angular clasts in cold regions are the result of freeze-thaw weathering and of the subsequent use of such interpretation as a basis for paleoclimatic reconstruction of Quaternary environments.

During the past decade, there has been an increase in laboratory experimentation which has supplied valuable information to improve the understanding on the behavior of, for example, carbonate rocks subjected to freeze-thaw and thermal shock (Yavuz et al. 2006). More recently, Schwarmborn et al. (2012) highlighted the role of freeze-thaw mechanical weathering by means of field and laboratory data. Laboratory testing, for example, showed that over 100 freeze and thaw cycles fractured quartz grains preferentially over feldspar. Microscopic features demonstrated that freeze-thaw cycling disrupted quartz grains along mineral impurities such as bubble trails, gas–liquid inclusions, or mineralogical subgrain boundaries. Single-grain micromorphology showed how it was defined by cryogenic fracturing and how it could serve as first-order proxy data for permafrost conditions in Quaternary records.

The mechanism of **hydration shattering** is based on the observation that unfrozen water at vey low temperatures is a significant factor in mechanical weathering. Adsorbed water (hydration) drawn in thin films between grains by *electro-osmosis* can generate a force of up to 2,000 kg cm^{-2}, sufficient to free grains and disintegrate rocks; hydration shattering is backed by experimental support and can be significant even without reaching freezing temperature (White 1976).

Laboratory studies have shown that the pressure exerted by **ice crystal growth** (polycystalline ice) from a melt reaches about 20 kPa. Nevertheless, certain conditions would have to be met for the process to work efficiently, such as a slow rate of freezing, which allows the flow of supercooled water through permeable materials to zones of crystallization (Fig. 2.3). Engineering studies have provided evidence based on the reaction of concrete to freezing: significant **hydraulic pressures** may develop in front of a freezing plane (e.g., Matsuoka and Murton 2008).

Fig. 2.3 Profusely fractured rocks in Antarctica, where ice crystal growth probably plays a significant role in the continued cracking action. Photograph by M.A.E. Chaparro taken at Vega Island (NE of Peninsula Antarctica)

2.3 Salt Weathering

Rock breakdown can also be caused by salts, which can be formed in nature (as it occurs in laboratories) from the reaction between acids and bases. The weathering action occurs either when they crystallize from solution or when, already as crystals, they expand on heating or hydration, as it happens with sodium sulfate ($Na_2SO_4 \rightarrow Na_2SO_4 \cdot 10H_2O$). Expanding salt crystals exert a pressure on the walls of the rock pores that exceeds the tensile strength of the rock. This process is known as *haloclasty*. The salt derives from an external source, such as capillary rising ground water, wind-blown with dust, sea water along rocky coasts, and atmospheric pollution.

Most salts intervening in weathering processes are water soluble, so that they are only widely found in arid or semi-arid areas, such as hot or cold deserts. Although other salts occur, the most common in hot deserts include $CaCO_3$, as well as chlorides, sulfates, and nitrates, of sodium and magnesium. In cold (arctic) deserts, the sulfates of sodium and magnesium are two of the most common salts.

Salts that frequently occur in deserts may be *anhydrous*, such as thenardite (Na_2SO_4), or anhydrite ($CaSO_4$); others may associate with water molecules, resulting in a *hydrate* compound, such as gypsum ($Na_2SO_4 \cdot 2H_2O$), mirabilite ($Na_2SO_4 \cdot 10H_2O$), kieserite ($MgSO_4 \cdot H_2O$), epsomite ($MgSO_4 \cdot 7H_2O$), among other examples.

Salts are an effective weathering agent if they enter rock pores, characteristically in solution. Research on the subject suggests that the ability of solutions to penetrate rock is connected with the salt involved. There is some evidence, for example, that a halite solution is absorbed faster and reaches a greater degree of saturation than a solution of anhydrite. It is also common along coasts. An example of salt weathering can be seen in the honeycombed stones in sea walls. A honeycomb is a type of *tafoni*, a class of cavernous rock weathering structure, which likely develops in large part by chemical and physical salt weathering processes (Fig. 2.4).

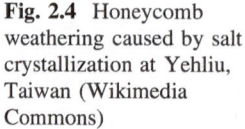

Fig. 2.4 Honeycomb weathering caused by salt crystallization at Yehliu, Taiwan (Wikimedia Commons)

Hydration, crystallization, and thermal expansion are additional types of expansionary force that are believed to cause rock failure.

Hydration pressure develops when salts hydrate and generate a considerable increase in volume. For example, when thenardite hydrates to mirabilite the resulting increase in volume exceeds 300 %. These sulfates may pass through the complete hydration–dehydration cycle in less than a day. Nonetheless, temperature must reach a transition point at which the less hydrated phase becomes fully hydrated; for example thenardite (Na_2SO_4) changes to mirabilite ($Na_2SO_4 \cdot 10H_2O$) at 32.4 °C. Generally, the greater the degree of hydration, the greater the pressure increase produced; pressures of up to 6.3 MPa have been recorded in laboratory studies whereas rock tensile strengths fluctuate between 2 and 20 MPa (Bland and Rolls 1998).

Crystallization pressure is associated with the evaporation of an aqueous solution in a rock cavity that leads to a concentration increase of dissolved phases. The aqueous solution may reach a concentration level close to saturation and, if evaporation continues, the crystallization process may slowly begin. If, however, the solution reaches supersaturation, crystallization may be deferred for some time and then rapidly occur. If the solution is cooled, as in deserts during night hours, the effect will depend on the concentration; a dissolved solids concentration near saturation may lead to crystallization.

Crystal growth gives rise to crystallization pressure. For instance, the crystallization pressure of halite, which is a common salt in dry and hot playa environments, is about 65 MPa when it is oversaturated by a factor of two; it increases to almost 220 MPa when it is oversaturated by a factor of ten, at 50 °C (Bland and Rolls 1998).

If the **thermal expansion** coefficient of a salt exceeds that of the surrounding rock, then the conditions are set for physical weathering to occur. For example, a temperature rise in a halite crystal of about 50 °C generates a volumetric expansion of 1 %, which is larger than that of any surrounding rock. Laboratory experiments that demonstrate this process are deemed insufficient and hydration and crystallization pressures are considered more important.

2.4 Insolation Weathering

Insolation (i.e., exposure to the sun) weathering is also known as thermal stress and is a consequence of expansion and/or contraction of rock determined by significant temperature range and rate of change (e.g., Weiss et al. 2004). For over 30 years, thermal fatigue of rocks has been investigated in dry as well as in wet conditions (e.g., Aires-Barros et al. 1975). In the Sahara desert, for example, the range may reach 42 °C over 24 h, whereas the rate of change is about 6 °C h^{-1} in the morning hours. The process appears as straightforward: a rock surface layer expands when it is heated by the sun, and then contracts when the heat source is momentarily discontinued. Cracking occurs when the stresses due to expansion and contraction surpass the rock's elastic limit. Recent research suggests that directional insolation may play a key role in the initial generation and propagation of meridional fractures (i.e., cracks with orientations not attributable to rock anisotropies or shape). These studies also indicate that rock size, latitude, and surface age exert an important influence with respect to their weight on rock fracture (Cary Eppes et al. 2010).

One aspect to consider is the source of energy that, in this case, is solar radiation, which is affected by a number of factors, fixed (such as latitude or altitude) or variable (such as cloud cover, wind speed, and rainfall). Therefore, temperature fluctuations may be short term and involve rapid change.

Another aspect to consider is rock *albedo*, which will influence the amount of effective solar heat received. The migration of this heat through a rock is a consequence of its thermal conductivity, which is generally low (e.g., a basalt has 0.96 and a granite 1.65 $Wm^{-1}K^{-1}$), implying relatively high surface temperatures, pronounced gradients (typically, between 0.5 and 1.2 °C cm^{-1}) and expansion of the surface layers. Upon cooling, contraction occurs and, hence, a change in volume is determined by the coefficient of thermal expansion, which also varies markedly among rock-forming minerals (Fig. 2.5). The repeated heating and

Fig. 2.5 Desert pavement at Laguna Brava (La Rioja, Argentina). Most clasts show sharp edges, a likely indication of thermally induced cracking and minimal transport. Photograph by M. A. Cioccale

Fig. 2.6 Exfoliation of granite dome rock in the enchanted rock state natural area, Texas, USA (Wikimedia Commons)

cooling exerts stress on the outer layers of rocks (i.e., a sort of *thermal fatigue*), which can cause their peeling off in thin sheets. The process of peeling off is also called *exfoliation*.

Thermal stress fatigue is not only associated with high temperatures in warm deserts. It appears as significant in cold regions as well. Thanks to high-frequency rock temperature data (i.e., temperature measurements at 1 min intervals), it has been possible to emphasize the significance of mechanical weathering in cold but dry regions, where rock disaggregation can be explained by thermal stress/shock (Hall 1999). For example, there is considerable argumentation over the significance of freeze-thaw as compared with thermal stress/shock. Studies on the dynamics of thermal stress performed in Antarctica have underlined the significance of the rate of change of temperature ($\Delta T/t$, where T is temperature and t is time) as significant factor, not only in the evaluation of freeze-thaw mechanism but mostly in thermal stress fatigue. The data gathered evidently showed that thermal stress fatigue and thermal shock may be more dynamic components of the Antarctic weathering regime than have generally been recognized. Clearly, the aridity of the area of study confines the role of freeze-thaw weathering. Values of $\Delta T/t$ of ≥ 2 °C min^{-1} suggest that thermal stress fatigue/shock is remarkably active were recorded; observations of rock flaking are thought to be the consequence of thermal stress (Hall and André 2001).

A process that may be associated with insolation effects is the so-called **dirt cracking**, which appears to be a wedging process started inside crevices by precipitation of laminar carbonate and then expanded by the wetting and drying of wind-blown expansive clays that exert pressure on fracture sides, widening the fissure (Dorn 2011).

2.5 Pressure-Release Weathering

When overlying materials (not necessarily rocks, because glacial retreat can also lead to this kind of mechanical weathering) are removed by erosion, for example, underlying rocks, originally subjected to a high confining pressure, may expand and fracture parallel to the surface, producing sheet or dilation joints (Fig. 2.6). These joint sets parallel to the local land surface have been widely reported in exhumed granitic terrains. This process is also known as exfoliation, and occurs over time, when sheets of rock break away from the exposed rocks along fractures. Sheet jointing related to a change in compressive stress is the best known pressure-release weathering process but sheet jointing involving mainly tensile stress has also been reported.

2.6 Wetting and Drying

In this weathering process, rocks are physically disintegrated by the accumulation of successive layers of adsorbed water molecules inside fractured rocks or mineral grains. This process is sometimes called slaking and occurs when polar water molecules are attracted to the walls of a fine crack. Consecutive cycles of wetting and drying will tend to give rise to expansion and contraction because the swelling pressure of water addition to the crack (i.e., strain during wet periods), may be followed in dry periods by attraction when residual water molecules on opposing faces of the crack are pulled together. Shale is particularly vulnerable to this process, which was also recorded in sandstone, limestone, schist, etc.

As wetting and drying proceeds, the size and/or number of microfissures and pores in rocks and minerals increases, thus contributing to the significance of both frost and salt weathering.

Glossary

Albedo: The reflectivity coefficient of a surface to short-wave radiation, i.e., the fraction of incoming solar radiation that is reflected rather than absorbed, from the Earth back into space. A value of 0 means the surface is a "perfect absorber" and a value of 1 means the surface is a "perfect reflector."

Anhydrous: A substance free from water, especially water of crystallization.

Biosphere: Part of the Earth's crust, waters, and atmosphere that supports life; the global sum of all ecosystem that interact within the elements of the lithosphere, hydrosphere, and atmosphere.

Electro-osmosis: a.k.a. electro osmotic flow, i.e., the motion of water or any other liquid induced by a potential across a porous material, capillary tube, or any other fluid conduit.

Exfoliation: Surface-parallel fracture systems in rock often leading to erosion of concentric slabs.

Haloclasty: A type of physical weathering caused by the growth of salt crystals. The process is first started when saline water seeps into cracks and evaporates depositing salt crystals, when the rocks are then heated, the crystals will expand putting pressure on the surrounding rock which will over time splinter the stone into fragments.

Hydrate: A substance that contains water of crystallization or water of hydration, in a definite ratio as an integral part of the crystal.

Lithosphere: Rigid outermost shell of a rocky planet. On Earth, it comprises the crust and the portion of the upper mantle. It continually interacts with the atmosphere and hydrosphere. It is broken into about a dozen separate, rigid blocks, or plates.

Regolith: Layer of loose heterogeneous material covering solid rock. It includes dust, soil, broken rock, and other related materials and is present on Earth, the Moon, Mars, some asteroids, and other terrestrial planets and moons.

Tafoni: Cave-like features found in granular rock with rounded entrances and smooth concave walls.

Thermal fatigue: The progressive and localized structural damage that occurs when a material is subjected to repeated heating and cooling. If the loads are above a certain threshold, microscopic cracks will begin to form at the stress concentrators such as the surface, persistent slip bands, and grain interfaces. Eventually a crack will reach a critical size, and the structure will suddenly fracture; materials do not recover when rested.

References

Aires-Barros L, Graça RC, Velez A (1975) Dry and wet laboratory tests and thermal fatigue or rocks. Eng Geol 9:249–255

Bland W, Rolls D (1998) Weathering: an introduction to the scientific principles. Arnold, London

Camuffo D (1995) Physical weathering of stones. Sci Total Environ 167:1–14

Cary Eppes M, McFadden LD, Wegmann KW et al (2010) Cracks in desert pavement rocks: further insights into mechanical weathering by directional insolation. Geomorphology 123:97–108

Dorn RI (2011) Revisting dirt cracking as a physical weathering process in warm deserts. Geomorphology 135:129–142

Hall K (1995) Freeze-thaw weathering: the cold region "Panacea". Polar Geogr 19(2):79–87

Hall K (1999) The role of thermal stress fatigue in the breakdown of rock in cold regions. Geomorphology 31:47–63

Hall K, André MF (2001) New insights into rock weathering from high-frequency rock temperature data: an Antarctic study of weathering by thermal stress. Geomorphology 41:23–35

Matsuoka N, Murton J (2008) Frost weathering: recent advances and future directions. Permafrost Periglac Process 19:195–210

Papida S, Murphy W, May E (2000) Enhacement of physical weathering of building stones by microbial populations. Int Biodeterior Biodegrad 46:305–317

Schwamborn G, Schirrmeister L, Frütsch F, Diekmann B (2012) Quartz weathering in freeze-thaw cycles: experiment and application to the El'gygytgyn Crater Lake record for tracing Siberian permafrost history. Geogr Ann A Phys Geogr 94(4):481–499

Scott KM, Pain CF (eds) (2009) Regolith science. Springer, Dordrecht

Wallder JS, Hallet B (1986) The physical basis of frost weathering: toward a more fundamental and unified perspective. Arct Alp Res 18(1):27–32

White SE (1976) Is frost action really only hidration shattering? A review. Artc Alp Res 8:1–6

Weber S, Gruber S, Girard L (2012) Design of measurement assembly to study in situ rock damage driven by freezing. In: Hinkel KM (ed) Proceedings of the 10th international conference on permafrost, vol 1. TICOP, Salekhard

Weiss T, Siegesmund S, Kirchner D et al (2004) Insolation weathering and hygric dilatation: two competitive factors in stone degradation. Environ Geol 46(3–4):4012–4413

Yavuz H, Altindag R, Sarac S et al (2006) Estimating the index properties of deteriorated carbonate rocks due to freeze-thaw and thermal shock weathering. Int J Rock Mech Min Sci 43:767–775

Chapter 3
The Biological Path to Rock Breakdown

Abstract Biological weathering is exerted through both, biophysical and biochemical corridors. Considering the recent scientific advances in applied microbiology, it is now possible to attain a more accurate view on the role of biology in the breakdown of minerals and rocks in the Earth's material cycle. Roots, lichens, mosses, algae, and bacteria are significant agents in mineral and rock breakdown, even exerting in some cases a comminuting action that promotes further the ensuing bio- or geochemical effect. Microorganisms tackle such action by producing aggressive substances (e.g., organic acids) that dissolve minerals and produce secondary solid and soluble phases which participate, through their riverine exportation to world oceans, in the process of continental denudation. The role of bacteria in dissolving metal sulfides, as in tailing impoundments resulting from mining operations, is particularly important to contribute to the understanding of the interaction of biota with the inorganic realm.

Keywords Biological weathering · Critical zone · Biophysical weathering · Biochemical weathering · Microbes · Bacteria · Lichens · Algae · Mosses · Fungi

3.1 Introduction

Weathering processes occur at the intersection of the biosphere with the remaining Earth's spheres: atmosphere, lithosphere, and hydrosphere. This is the zone where rock meets life and it is increasingly known as the *critical zone* (e.g., Anderson et al. 2007). Therefore, the role of living organisms in the process of material breakdown is so characteristic that it must be recognized as a separate category, identifying biophysical and biochemical processes as accepted subdivisions. Decaying organic matter, plant roots, mosses, microbes (algae, fungi, lichens, and bacteria) are biological agents that actively participate in mineral and rock collapse (Fig. 3.1).

P. J. Depetris et al., *Weathering and the Riverine Denudation of Continents*, SpringerBriefs in Earth System Sciences, DOI: 10.1007/978-94-007-7717-0_3,

Fig. 3.1 Rocks profusely colonized by lichens and moss at Wulaia Bay, Navarino Island (Tierra del Fuego, Chile, ~55° S 68° W). Photograph by P.J. Depetris

Two main pathways are used by living matter to support mineral and rock breakdown: they may apply physical stress through, for instance, the expansion and contraction of plants or roots, or they may produce substances such as carbon dioxide, several acids (inorganic or organic), complexing agents, protons, and electrons, all of which are part of their life processes.

3.2 Biophysical Weathering

Seedlings sprouting in crevices and plant roots may exert physical pressure as well as providing a pathway for water infiltration. Frequently expressed ideas imply that, as roots grow, they press against the rock and put stress on the joints they are growing in. Over time, this stress breaks rocks apart. Some authors, however, argue that the **effect of roots** is far from efficient in as much as effective tensile stress is set up by radial pressure which, in the case of roots is only 1/3–1/4 of the axial amount, typically of the order of three MPa, only sufficient to fracture some weak rocks (Bland and Rolls 1998). Other authors have demonstrated that some tracks and holes in minerals that have been mistakenly attributed to the action of roots were, in fact, produced by chemical dissolution mechanisms (Sverdrup 2009). At any rate, the mechanical role of roots appears as still open to discussion although visual evidence appears to support their significance (Fig. 3.2).

The mechanical effect of roots seems to be more effective in regolith, soils, and unconsolidated sediments. Li et al. (2006) have shown that in China's loess plateau, plant roots have stronger effects on soil physical properties than on chemical characteristics; the role of plant roots in controlling soil weathering and leaching increases in the following order: infiltration enhancement > increased flow of bioactive substances > stabilization of soil structure.

Fig. 3.2 Trees growing in schists and phyllites from "Pizarroso-Cuarcítico" Devonian Group in Sotiel Coronada, Huelva, Spain. Photograph by K.L. Lecomte

Lichens are composite organisms, consisting of fungi and algae harmonically growing together (e.g., Purvis 2000). Lichens are capable of physically breaking down rocks, by the penetration of hyphae (whose growth-induced stress may exceed the tensile strength of most rocks), and by increasing their water content (typically between 150 and 300 %), thus reaching damaging levels of tensile stress by volume increase (Chen et al. 2000). The physical effects of lichens are also reflected by the mechanical disruption of rocks caused by expansion and contraction of lichen thallus, and the swelling action of organic and inorganic byproducts originating from lichen activities (Chen et al. 2000). Lee and Parsons (1999) have shown the significant weathering effect of the crustose lichen *Rhizocarpon geographicum* by both, biomechanical and biochemical means, on a Lower Devonian granite. Biomechanical weathering is mediated by fungal hyphae that enter the rock via mineral boundaries at a rate $\geq$0.002–0.003 mm y^{-1}. Once inside the rock, hyphae make use of intragranular pores along weak planes in biotite, alkali, and plagioclase feldspar. Biotite grains exposed at the lichen–granite interface have been fragmented by biomechanical action in less than 122 years. After extensive biomechanical weathering (e.g., outcrop surfaces exposed for ~10 Kyr), sub-mm sized fragments of biotite (which show the clearest evidence for biochemical weathering) and plagioclase feldspar abound in the lower parts of the lichen's thallus (Lee and Parsons 1999). The relative significance of weathering by lichens appears as substantial in extremely cold environments, like in Antarctica (Fig. 3.3), where they act mechanically and chemically (Jie and Blume 2002).

Fig. 3.3 **a** Lichens growing on pebbles and cobbles in a large ice-free area in the James Ross Island (Antarctica's NW). **b** Detail of **a**. Photographs by K.L. Lecomte

The effectiveness of **algae** as agents of mechanical weathering also shows significance in cold climates. Studies have revealed that the algal mucilage that surrounds algal cells would incorporate a large quantity of water during wet periods (e.g., 20 times the volume of the dry state), thus producing an expansionary force sufficient to separate already-weakened rock pieces. Observations performed on a number of *nunataks* of the Juneau Icefield in Alaska indicate that algae play a significant role in the breakdown of granitic rock. Data gathered at Juneau by Hall and Otte (1990) suggested that the average mass of material loss could be as high as 562 g m^{-2} yr^{-1}.

To conclude with the biophysical aspects of weathering, it is worth mentioning that an experimental approach has shown that **microbes** promote physical weathering. The use of a combination of physical and biological processes in several experiments with building stones has shown that the extent of decay in limestone was significantly enhanced when compared with the physical or biological agents acting alone (Papida et al. 2000).

3.3 Biochemical Weathering

The zone dominated by plant roots is referred to as the *rhizosphere* (probably the most effective weathering microenvironment), where **roots** may bring about material breakdown, particularly through biochemical paths. Roots are systems that, as part of a complex life-sustaining process, emit and absorb components. The tip of the main root releases organic acids of low-molecular-weight (e.g., citric, lactic, and tartaric acid), along with protons and electrons, which contribute to mineral decay through the delivery and mobilization of iron and aluminum in chelate-like compounds. Experimental work using polished marble tablets has shown that root etching occurs across grains and at grain boundaries, and emerges as more severe at the junction of root networks (Mottershead et al. 2003). More recently, Calvaruso et al. (2006) have shown that on the basis of column experiments with a quartz-biotite substrate, for example, pine roots significantly increased biotite weathering by a factor of 1.3 for magnesium and 1.7 for potassium. They demonstrated as well that the inoculation of *Burkholderia glathei* significantly increased biotite weathering by a factor of 1.4 for magnesium and 1.5 for potassium in comparison with the pine alone.

In the Ouachita Mountains (Arkansas, USA.), Phillips et al. (2008) gathered evidence on the rapid initial rate (5–10 mm yr^{-1}) of soil production, which is facilitated by aggressive vegetation colonization, along with a favorable regional climate and sediment accumulation. Plant establishment (i.e., it takes <10 yr of pedogenic maturation in the case of trees) accelerates local weathering rates.

After the stage of physical weathering (i.e., disaggregation and fragmentation) of the rock in immediate contact with **lichens**, the ensuing chemical weathering is fundamentally due to the excretion of organic acids. As a result, many rock-forming minerals exhibit extensive surface corrosion. Moreover, the specific literature points out that oxalic acid and so-called "lichen acids" (e.g., physolic and lobaric acids) are considered biomolecules particularly active as weathering agents; chelation of metallic cations is also important (e.g., Chen et al. 2000). In regions with extreme climate, such as the Qinghai-Tibet Plateau, the gathered evidence indicated that the assumed dominance of physical weathering processes on granite, such as freeze–thaw was in fact, attributable to the action of lichens (Hall et al. 2005). It is also important to underline that the effectiveness of lichens as biochemical weathering agents appears as significant on different rock substrates. Besides granite, lichens have also proved successful as weathering agents on Lanzarote's lava flows (Stretch and Viles 2002), on a mica-schist mini-watershed (<1 m^2) (Aghamiri and Schwartzman 2002), and on a Calabrian granodiorite from the Sila uplands (Scarciglia et al. 2012). Chen et al. (2000) have produced an interesting review, mainly on the effect of lichens on different rock substrates and on the weathering byproducts that they generate. Almost concurrent, Adamo and Violante (2000) published an extensive review contributing to the knowledge on the characteristics of lichen-induced weathering and, also, on the likely formation of oxalate/"lichen acid"-derived minerals.

The role of **fungi** in the weathering processes occurring in forest ecosystems is also well-established, where biological activity in general and fungi in particular are significant drivers of organic matter decomposition, cycling of nutrients, mineral formation, and metal dynamics. The mineral soil horizons of boreal forest are intensively colonized by mycorrhizal mycelia (forming symbioses with forest trees) which

Fig. 3.4 **a** Antarctic volcanic rocks blanketed by mosses (James Ross Island). **b** Blown up detail of **a**. Photograph by K.L. Lecomte

transfer organic metabolites and protons to mineral surfaces, resulting in mineral dissolution and mobilization of nutrients and metal cations (Finlay et al. 2009).

Although the weathering effect of **mosses** has been less investigated than the one exerted by lichens, algae, or fungi, it is now clear that their function in dissolving rocks is significant nowadays and was even more important in the geological past. Lenton et al. (2012), for example, probed into their role during the Ordovician, when nonvascular plants appear to have affected the weathering rate and, hence, climate, significantly lowering ambient temperature. The weathering "amplification factor" due to the presence of moss was significant for Al, Ca, Fe, K, and Mg from silicates. The spreading out of nonvascular land plants during the Ordovician accelerated chemical weathering and may have drawn down enough atmospheric CO_2 to activate the growth of ice sheets. There are recent case studies that have looked into the current role of nonvascular plants and lichens in their relative significance in the control of weathering in cold climates (e.g., Zakharova et al. 2007; Guglielmin et al. 2012). Figure 3.4 shows mosses growing on volcanic rocks in Antarctica peninsula.

Bacteria are uni- or multicellular microscopic organisms that have been active in the Earth's surface for perhaps 3,100 million years are found in both, aerobic and anaerobic environments. They may be *heterotrophic* (i.e., feed on organic molecules), *autotrophic* (i.e., are sustained by simple inorganic compounds), or *mixotrophic* (i.e., use both nutrient sources). In biologically active soils, CO_2 may be concentrated by 10–100 times the amount expected from equilibrium with atmospheric CO_2 (i.e., almost 0.004 %), yielding H_2CO_3 and H^+ via dissociation. The result is a lowering of the pH of soil solutions (i.e., typically to 4 or 5) and the ensuing increase of weathering reactions through a biochemical pathway. In the following equations, organic matter is represented by the generalized formula for carbohydrate, CH_2O.

$$CH_2O_{(s)} + O_{2(g)} \rightarrow CO_{2(g)} + H_2O \tag{3.1}$$

The respiration of soil organic matter has, as it will be seen in the next chapter, a major impact on the chemical weathering of minerals.

Nitrification is a substantial part of a larger chemical sequence known as the *nitrogen cycle*, and it is the general term to describe the conversion of ammonium ions (products of the bacterial processing of dead organic matter) to nitrites and nitrates. The significance of this process in biochemical weathering is that it involves the release of protons:

$$NH^+_{4(aq)} + 2O_{2(g)} \rightarrow NO^-_{3(aq)} + 2H^+_{(aq)} + H_2O \tag{3.2}$$

As a result of this bacteria-mediated reaction, two hydrogen ions are produced for every ammonium ion and acidification occurs, thus affecting the weathering rate.

Several laboratory experiments have been carried out to prove the occurrence of biologically-induced mineral dissolution. Ullman et al. (1996) obtained experimental evidence for silicate minerals dissolution mediated by microbes, proving that some microbial metabolites enhanced dissolution rates by a factor of ten above the expected nonbiological alteration. Barker et al. (1998) reported the increased release of cations from biotite (Si, Fe, Al) and plagioclase (Si, Al) by up to two orders of magnitude by microbial activity compared to abiotic controls. Song et al. (2007) analyzed the effect of *Bacillus subtilis* on granitic rocks, concluding that bacteria have a strong influence in the granite weathering by forming pits at as rate 2.4 times faster than bacteria-free specimen.

The role of bacteria in biochemical weathering may be emphasized further by mentioning their participation in the oxidation of metals, as it happens with sulfides when they are transformed into water-soluble metal sulfates. Oxidation happens at a slow rate in very acid waters; below pH 3.5, iron oxidation is catalyzed by the iron bacterium *Thiobaccillus thiooxidans.* At pH 3.5–4.5 oxidation is catalyzed by *Metallogenium.* Clearly, acid conditions foster microbial alteration raising significantly their reaction rates when compared to nonbiological mechanisms. Figure 3.5 shows an acidic environment-type and the associated colonies of acidophilus bacteria and algae.

The oxidation of iron pyrite to ferric sulfate is an example of the above mentioned mechanism:

Fig. 3.5 **a** Extreme acidic environment in Amarillo River (La Rioja, Argentina, ~29° S ~67° 30′ W). **b** Associated colonies of acidophilus algae and bacteria. Photographs by K.L. Lecomte

$$FeS_{2(S)} + 3.5O_{2(g)} + H_2O \rightarrow Fe^{2+}_{(aq)} + 2H^{+}_{(aq)} + 2SO^{2-}_{4(aq)} \quad (3.3)$$

$$Fe^{2+}_{(aq)} + 0.5O_{2(g)} + 2H^{+} \rightarrow Fe^{3+}_{(aq)} + H_2O \quad (3.4)$$

Bacteria use iron compounds to obtain energy for their metabolic demand [i.e., such as the oxidation of Fe(II) to Fe(III)]. Since these bacteria derive their energy from the oxidation of inorganic matter, they thrive where organic matter is rare or missing, using CO_2 as a source of carbon. However, energy is not obtained efficiently from iron oxidation and approximately 220 g of Fe^{2+} must be oxidized to produce 1 g of carbon. It is not surprising then that large deposits of Fe(III) oxide develop in areas where iron-oxidizing bacteria endure.

Investigations in acid mine drainage have produced a wealth of information on the interaction of metal sulfides and bacteria, where their relationships may be complex; Ehrlich (1996) reported several satellite microorganisms that live in close association with *Acidithiobacillus ferrooxidans*. More recent data (e.g.,

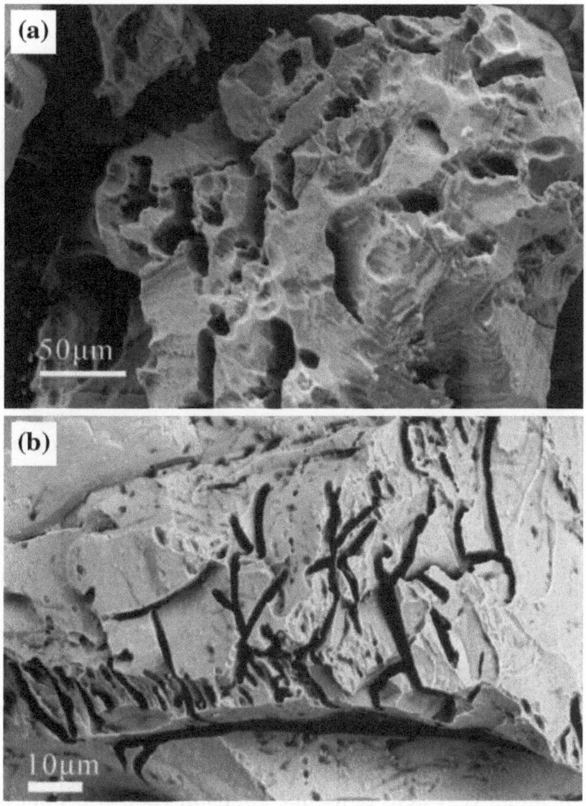

Fig. 3.6 Scanning electron micrographs of corrosion etches on sulfide minerals caused by surface-attached *Acidithiobacillus ferrooxidans* cells. **a** Arsenopyrite particles bio-oxidized for 15 days. **b** Broken surface of pyrite particles bio-oxidized for 36 days (Lu and Wang 2012). Reproduced with permission, Mineralogical Society of America

Halinen et al. 2009; Ziegler et al. 2009) show complex communities structures in pyrite oxidation and bioleaching operation (Fig. 3.6). It is nowadays recognized that complex ecological interactions control the biogeochemical element cycles in acid environments, like in the Tinto River, in Spain (G-Toril et al. 2003). *Leptospirilum ferrooxidans* (Edwards et al. 1998; Diaby et al. 2007; Rawlings and Johnson 2007), heterotrophic bacteria, green algae, fungi, yeasts, mycoplasma, and amoebae have all been reported from acid mine waters.

The model of Fig. 3.7 sums up the observed changes within tailing impoundments proposed by Diaby et al. (2007). As the water level inside the impoundment decreases, oxygen promotes the oxidative dissolution of sulfide minerals (e.g., pyrite), chiefly by *Leptospirillum* spp. acidophiles (*Sulfobacillus* spp. and *At. ferrooxidans*) add to this process by generating H_2SO_4, as well as by oxidizing Fe(II). *Lysates* and exudates (mainly dissolved organic carbon or DOC that leaches out from the pores of injured tissue) from autotrophic acidophilus sustain the growth of heterotrophic acidophilus (iron-reducing *Acidiphilium* and *Acidobacterium*-like bacteria, and sulfate-reducing *prokaryotes*) while oxidation products of the primary producers [Fe(III) and sulfate] act as terminal electron acceptors for the heterotrophs where oxygen is limiting or absent. Below the oxidation front,

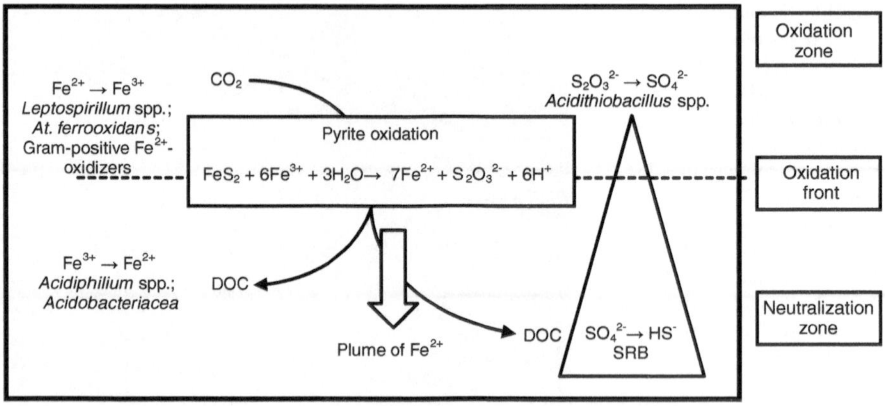

Fig. 3.7 Model of the microbial impact on geochemical dynamics observed at Piuquenes tailings impoundment. *DOC* dissolved organic carbon. *SRB* sulfate-reducing bacteria (Diaby et al. 2007). Reproduced with permission, John Wiley and Sons

dissimilatory reduction of Fe(III) and sulfate is considered to be the leading geochemical processes, both of which are eventually limited by the availability of these oxidized species or by electron donors. The oxidation front will continue to migrate downwards, depending on the rate at which the water table falls. In due course, if and when the impoundment is completely drained, the mineral debris could become essentially fully oxidized, resulting in the dissolution of significant quantities of waste sulfide minerals from the mining operation and the generation of metal-rich acid effluents. Johnson (2009) produced an excellent review on the nature of extremely acid environments and on the biodiversity of microorganisms found within them. At any rate, all these biochemical mechanisms end up playing a role in continental wearing down.

The influence of microorganisms persists throughout the occurrence of regolith processes that affect all the materials that will be eventually removed from the continents. Reith et al. (2009) have produced a comprehensive review on the impact of microorganisms on regolith that should be examined by all those interested in probing into the chain of processes that intervene in continental denudation.

Glossary

Autotrophic: Any organism capable of self-nourishment by using inorganic materials as a source of nutrients and using photosynthesis (photoautotrophs) or chemosynthesis (lithoautotrophs) as a source of energy as most plants and certain bacteria and protists. Autotrophs can reduce carbon dioxide to make organic compounds, creating a store of chemical energy.

Chelation: A particular way that ions and molecules bind metal ions.

Critical zone: System of coupled chemical, biological, physical, and geological processes operating together to support life at the Earth's surface.

Heterotrophic: An organism that cannot synthesize its own food and is dependent on complex organic substances for nutrition. This contrasts with autotrophs, such as plants and algae.

Lysates: Solutions produced when cells are destroyed by disrupting their cell membranes.

Mixotrophic: Any organism capable of existing as either an autotroph or heterotroph. Mixotrophs can be either eukaryotic or prokaryotic.

Nitrification: The substitution of a nitro group for another group in an organic compound. The biological oxidation of ammonia with oxygen into nitrite followed by the oxidation of these nitrites into nitrates.

Nitrogen cycle: Continuous sequence of events by which atmospheric nitrogen and nitrogenous compounds in the soil are converted, as by nitrification and nitrogen fixation into substances that can be utilized by green plants. Then substances return to the air and soils as a result of the decay of plants and denitrification. This transformation can be carried out through both biological and physical processes.

Nunataks: Areas of rock emerging above ice sheets and glaciers.

Prokaryotes: A group of organisms whose cells lack a membrane-bound nucleus (karyon).

Rhizosphere: Is the narrow region of soil that is directly influenced by root secretions and associated soil microorganisms.

References

Adamo P, Violante P (2000) Weathering of rocks and neogenesis of minerals associated with lichen activity. Appl Clay Sci 16:229–256

Aghamiri R, Schwartzman DW (2002) Weathering rates of bedrock by lichens: a mini watershed study. Chem Geol 188:249–259

Anderson SP, von Blackenburg F, White AF (2007) Physical and chemical control on the critical zone. Elements 3:315–319

Barker WW, Welch SA, Chu S et al (1998) Experimental observations of the effects of bacteria on aluminosilicate weathering. Am Mineral 83:1551–1563

Bland W, Rolls D (1998) Weathering: an introduction to the scientific principles. Arnold, London

Calvaruso C, Turpault M-P, Frey-Klett P (2006) Root-associated bacteria contribute to mineral weathering and to mineral nutrition in trees: a budgeting analysis. Appl Environ Microbiol 72(2):1258–1266

Chen J, Blume HP, Beyer L (2000) Weathering of rocks induced by lichen colonization—a review. Catena 39:121–146

Diaby N, Dold B, Pfeifer H-R et al (2007) Microbial communities in a porphyry copper tailings impoundment and their impact on the geochemical dynamics of the mine waste. Environ Microbiol 9(2):298–307

Edwards KJ, Schrenk MO, Hamers R et al (1998) Microbial oxidation of pyrite: experiments using microorganisms from extreme acidic environment. Am Mineral 83:1444–1453

Ehrlich HL (1996) Geomicrobiology. Dekker, New York

Finlay R, Wallander H, Smits M et al (2009) The role of fungi in biogenic weathering in boreal forest soils. Fungal Biol Rev 23:101–106

G-Toril E, L-Brossa E, Casamayor EO et al (2003) Microbial ecology of an extreme acid environment: the Rio Tinto river. Appl Environ Microbiol 69:4853–4865

Guglielmin M, Worland MR, Convey P et al (2012) Schmidt Hammer studies in the maritime Antarctic: application to dating Holocene deglaciation and estimating the effects of macrolichens on rock weathering. Geomorphology 155–156:34–44

Halinen AK, Rahunen N, Kaksonen AH et al (2009) Heap bioleaching of a complex sulfide ore. Part I: effect of pH on metal extraction and microbial composition in pH controlled columns. Hydrometallurgy 98(1–2):92–100

Hall K, Otte W (1990) A note on biological weathering on nunataks of the Juneau Icefield, Alaska. Permafrost Periglac Process 1:189–196

Hall K, Arocena JM, Boelhouwers J et al (2005) The influence of aspect on the biological weathering of granites: observations from the Kunlun Mountains, China. Geomorphology 67:171–188

Jie C, Blume HP (2002) Rock-weathering by lichens in Antarctic: patterns and mechanisms. J Geo Sci 12(4):387–396

Johnson DB (2009) Extremophiles: acidic environments. In: Schaechter M (ed) Encyclopedia of microbiology. Elsevier, Amsterdam

Lee MR, Parsons I (1999) Biomechanical and biochemical weathering of lichen-encrusted granite: textural controls on organic–mineral interactions and deposition of silica-rich layers. Chem Geol 161(4):385–397

Lenton TM, Crouch M, Johnson M et al (2012) First plants cooled the Ordovician. Nat Geosci 5:86–89

Li Y, Zhang Q, Wan G et al (2006) Physical mechanisms of plant roots affecting weathering and leaching of loess soil. Sci China Ser D Earth Sci 49(9):1002–1008

Lu X, Wang H (2012) Microbial oxidation of sulfide tailing and the environmental consequences. Elements 8(2):119–124

Mottershead DN, Baily B, Collier P et al (2003) Identification and quantification of weathering by plant roots. Build Environ 38:1235–1241

Papida S, Murphy W, May E (2000) Enhacement of physical weathering of building stones by microbial populations. Int Biodeterior Biodegrad 46:305–317

Phillips JD, Turkington AV, Marion DA (2008) Weathering and vegetation effects in early stages of soil formation. Catena 72:21–28

Purvis W (2000) Lichens. Smithsonian, Washington

Rawlings DE, Johnson DB (2007) Biomining. Springer, Berlin

Reith F, Dürr M, Welch S et al (2009) Geomicrobiology of regolith. In: Scott KM, Pain CF (eds) Regolith science. Springer, Dordrecht

Scarciglia F, Saporito N, La Russa ML et al (2012) Role of lichens in weathering of granodiorite in the Sila uplands (Calabria, southern Italy). Sed Geol. doi:10.1016/j.sedgeo.2012.05.018

Song W, Ogawa N, Oguchi CT et al (2007) Effect of *Bacillus subtilis* on granite weathering: a laboratory experiment. Catena 70(3):275–281

Stretch RC, Viles HA (2002) The nature and rate of weathering by lichens on lava flows on Lanzarote. Geomorphology 47:87–94

Sverdrup H (2009) Chemical weathering of soils and minerals and the role of biological processes. Fungal Biol Rev 23:94–100

Ullman WJ, Kirchman DL, Welch SA et al (1996) Laboratory evidence for microbially mediated silicate mineral dissolution in nature. Chem Geol 132: 11–17

Zakharova EA, Pokrovsky OS, Dupré B et al (2007) Chemical weathering of silicate rocks in Karelia region and Kola Peninsula, NW Russia: assessing the effect of rock composition, wetlands and vegetation. Chem Geol 242:255–277

Ziegler S, Ackermann S, Majzlan J et al (2009) Matrix composition and community structure analysis of a novel bacterial pyrite leaching community. Environ Microbiol 11(9):2329–2338

Chapter 4
Chemical Weathering Processes on the Earth's Surface

Abstract The exhumation of rocks from the Earth's crust implies that they must adjust thermodynamically to the conditions existing on the surface, which are extremely different from those prevailing during their formation, with higher temperature, pressure, and often exposed to chemically aggressive fluids. The processes involved in chemical weathering, such as dissolution, hydrolysis, etc., are at the core of the adjustment mechanism, transforming solid, and usually refractory rock material, into particles-typically stripped from part of their original components and dissolved phases, both of which are amenable to be transported from the continents to the sea. Although anthropogenic actions have altered natural denudation rates, still a relatively minor portion of the material thus produced stays for a longer period on the continents, temporarily sequestered in depositional systems. The most important participants in the weathering scenario are mineral dissolution, silicate hydrolysis, and redox reactions.

Keywords Dissolution · Hydration · Carbonation · Hydrolysis · Redox processes · Oxidation · Reduction · Cation exchange · Regolith · Isotopes

4.1 Introduction

As seen in a previous chapter, physical weathering is a mechanical process which fragments rocks into smaller particles without a substantial change in chemical composition; in contrast, biological weathering exerts jointly a biophysical and biochemical effect. As an example of the synergism that dominates in rock weathering, these processes are very important in the exogenous cycle since they vastly increase the surface area of crustal material exposed to the agents of chemical weathering, i.e., air and water. A larger surface area means that more mineral matter will be accessible to react with fluids (Fig. 4.1). All weathering processes can be viewed as the adjustment of rocks and minerals formed at high temperature and pressure contrasting with the situation prevailing on the Earth's surface, where low temperatures and pressure prevails (e.g., Anderson et al. 2007).

P. J. Depetris et al., *Weathering and the Riverine Denudation of Continents*, SpringerBriefs in Earth System Sciences, DOI: 10.1007/978-94-007-7717-0_4,

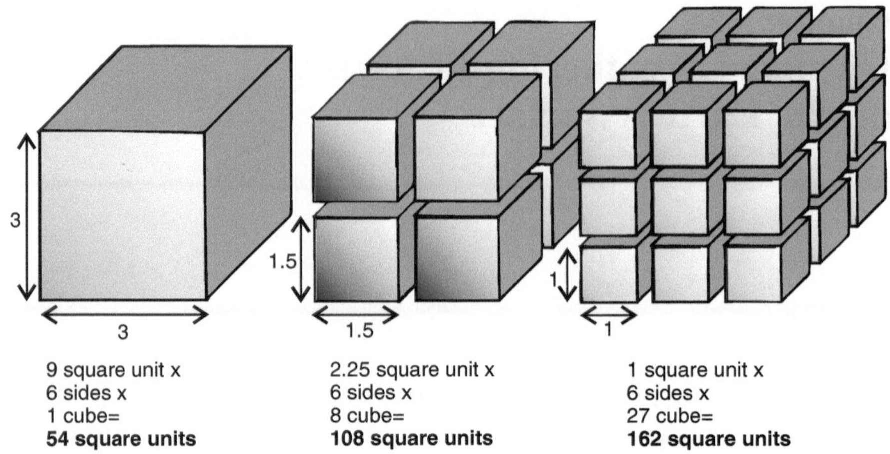

Fig. 4.1 The specific surface area (SSA) is the total surface area of a material per unit of mass, for example, $m^2\ g^{-1}$. The figure illustrates the increase of the surface area as the size of particles decreases

Therefore, minerals adjust to the new set of conditions to recuperate stability; this adjustment may be relatively fast for a Ca-rich plagioclase, for example, or slow as in the weathering of some silicates, like alkali feldspar or muscovite.

Among the many factors that control rock weathering such as climate, relief, biota, time, and the properties of the rock itself, climate plays a major role determining water availability and reaction kinetics which jointly determine the extent of chemical leaching. A clear example was presented by Benedetti et al. (2003), which studied the response of weathering in a region exposed to extremely high precipitation (1.8 to 12 m yr^{-1}). Clearly, chemical weathering is set off by water (mostly water with slightly acid pH) and gases (e.g., oxygen), which attack minerals (e.g., Drever 2005). Some compounds and ions of the original mineral are removed in solution, percolating through mineral debris to feed groundwater, rivers, and lakes. A proportion of fine-grained solids may be washed away from the weathering site, leaving a modified solid residue (i.e., *regolith*) which forms the basis of soils.

4.2 Mechanisms of Chemical Weathering

Chemical weathering is essentially an intricate process not only because of the complexity of the chemistry involved—some of which remains unknown—but also because minerals react differently to the aggression of weathering agents (e.g., White and Brantley 1995). Diverse mechanisms of chemical weathering are recognized and a variety of combinations occur together during the breakdown of the majority of minerals and rocks.

4.2.1 Dissolution

The most straightforward weathering reaction is the **dissolution** of soluble minerals. The water molecule is efficient in separating ionic bonds, such as in the dissolution of halite that results in an electrolyte solution:

$$NaCl_{(s)} \leftrightarrow Na^{+}_{(aq)} + Cl^{-}_{(aq)} \tag{4.1}$$

Halite's *solubility product* is $K_{sp} = 37.6$.

Clearly, halite is highly soluble; at 25 °C and a pressure of 100 kPa, its solubility is 6.15 mol kg^{-1} ($\sim$351 g L^{-1}). Since no hydrogen ions (H^{+}) are involved, the process is independent of pH. Due to a number of factors, like the remarkable abundance in seawater of the two ions involved, several authors have probed into the chemical characteristics of this simple mineral. For example, Gavrieli et al. (1989) studied the solubility of halite as a function of temperature in the highly saline brine system of the Dead Sea, and Alkattan et al. (1997) experimented on its dissolution kinetics.

Similarly, anhydrite, which is markedly less soluble than halite (i.e., its solubility is 0.67 g L^{-1}), is subjected to *congruent dissolution* when in contact with water;

$$CaSO_{4(s)} \leftrightarrow Ca^{2+}_{(aq)} + SO^{2-}_{4\,(aq)} \tag{4.2}$$

In contrast with halite, anhydrite's $K_{sp} = 2.4\ 10^{-5}$.

In some cases, the formation of a *hydrate* occurs with a definite number of molecules of water of crystallization;

$$CaSO_{4(s)} + 2H_2O \leftrightarrow CaSO_4 \cdot 2H_2O_{(s)} \tag{4.3}$$

4.2.2 Hydration

The **hydration** of anhydrite to gypsum results in "bubbles," often discernible in *playas*, as a consequence of expansion and mechanical deformation. Anhydrite and gypsum, as well as halite, are typical minerals of saline lakes and playas, where they crystallize as a consequence of evaporation, following a sequence that is governed by their solubility product (Fig. 4.2). Moreover, these minerals, abundant in arid environments, are easily windblown and transported away from their source, thus contributing significantly to the chemical composition of raindrops. Likewise, ocean aerosols determine that most atmospheric precipitation is of the Na^{+}-Cl^{-}- type. As rainfall moves away from the sea coast, the continental chemical signature becomes more prominent (e.g., Forti Adolpho et al. 2012) due to the participation of windblown dust and soluble aerosols. It follows, then, that a

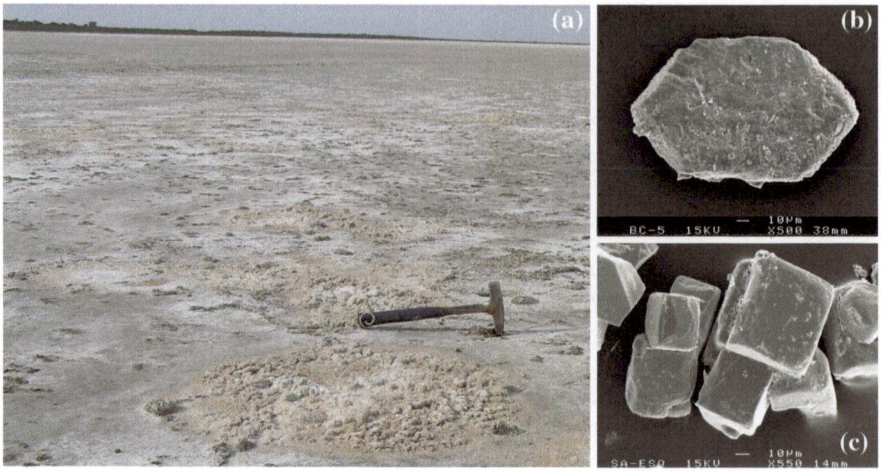

Fig. 4.2 **a** Typical playa basin (Salina de Ambargasta, ~29° S ~64° W), in the NW corner of Córdoba Province (Argentina), photographed during the desiccation stage (i.e., austral winter); the structure *left* by microbial mats can be clearly observed. **b** Pseudo-hexagonal gypsum crystal from Barranca Colorada (Salina de Ambargasta) obtained at 5 cm depth. **c** Cubic halite crystals from the surface salt crust at the playa basin of Salina de Ambargasta (Córdoba, Argentina) (Zanor et al. (2013). Photographs by G.A. Zanor

significant part of the dissolved phases in atmospheric precipitation are, in fact, recycled salts that originate in the sea and return to the sea after flowing over and through the outcropping continental crust, as rivers and aquifers. Many authors that have examined chemical denudation employed a methodology that allows correcting the contribution of chemical weathering, by subtracting the atmospheric input (e.g., Stallard and Edmond 1987; Gaillardet et al. 1999).

Silica is very abundant in the Earth's crust and, although it occurs naturally as eight distinct forms with uneven abundance, the species of real interest are ordinary quartz, opal-A, and amorphous silica (e.g., Wray 1997). Amorphous silica has a solubility of 60–80 mg L^{-1} at 0 °C, 100–140 mg L^{-1} at 25 °C, and about 300–380 mg L^{-1} at 90 °C (Krauskopf 1956; Krauskopf and Bird 1995). Clearly, silica solubility increases markedly with temperature. After its congruent dissolution, $Si(OH)°_4$ is the most representative *monomer* and usually exists as the uncharged acid, H_4SiO_4 (Fig. 4.3). Wray (1997) has completed an extensive review on solutional weathering in sandstones, and Piccini and Mecchia (2009) have studied the solution weathering rate and the origin of *karst* landforms and caves in quartzites of Venezuela.

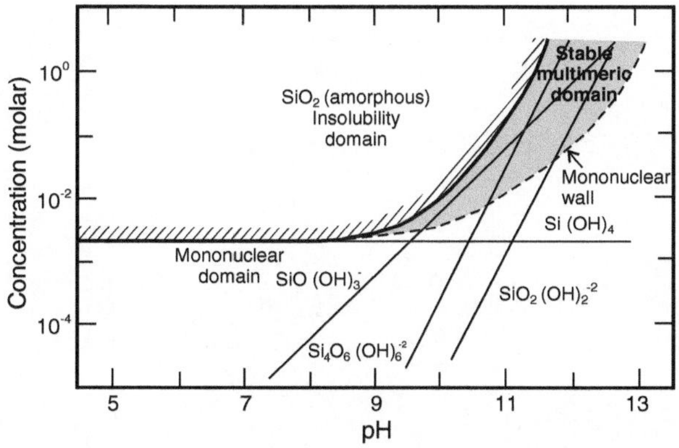

Fig. 4.3 Species in equilibrium with amorphous silica. In natural waters dissolved silica is present as monomeric silicic acid. The line surrounding the shaded area gives the concentration of maximum soluble silica. The mononuclear wall represents the lower concentration limit below which multinuclear silica species are not stable (Stumm and Morgan 1996). Reproduced with permission, John Wiley and Sons

4.2.3 Carbonation

Carbonation is perhaps the Earth's commonest geochemical mechanism and is concerned with the weathering effects of CO_2 in aqueous solution and, particularly, the interaction with $CaCO_3$ that in some parts may give rise to peculiar karstic landscapes. Water in equilibrium with the atmosphere has a pH of ~5.6 due to presence of CO_2 but groundwater may be more acidic as the concentration of CO_2 can be 20 to 30 times higher than in the atmosphere due to the production of CO_2 by biological respiration and the resulting higher partial pressure in soil air. Microbes, moss, and fungi are significant producers of CO_2 (see Chap. 3) and plants respire about 40 % of their CO_2 through the root system. In tropical soils, the interstitial air may contain up to 11 % of CO_2 due to the exuberant biological activity.

The following equation is of significant importance in chemical weathering because it produces carbonic acid:

$$CO_{2(g)} + H_2O \leftrightarrow H_2CO_{3(aq)} \tag{4.4}$$

The weak carbonic acid can dissociate further:

$$H_2CO_{3(aq)} \leftrightarrow H^+_{(aq)} + HCO^-_{3\,(aq)} \tag{4.5}$$

Bicarbonate ion itself dissociates also

$$HCO^-_{3\,(aq)} \leftrightarrow H^+_{(aq)} + CO^-_{3\,(aq)} \tag{4.6}$$

Fig. 4.4 Karstic "cenotes" in the Yucatán Peninsula (Mexico). The term derives from low-land Maya and refers to any location with accessible groundwater. It is a deep natural sinkhole that results from the collapse of limestone bedrock. In the Peninsula's N and NW, cenotes may be up to 100 m deep, with the water flow being more likely dominated by aquifer matrix and fracture flows. Conversely, the cenotes along the Caribbean coast often provide access to extensive underwater cave systems. **a** The so-called "sacred cenote" at the archeological area of Chichen Itzá (Wikimedia Commons). **b** A typical cenote (Chan Sinicché) at Cuzamá, on the eastern side of Yucatán (Wikimedia Commons)

Equations 4.5 and 4.6 clearly show the dependency of the system with the pH of the solution, as Bjerrum plots clearly show (e.g., Drever 1997). Moreover, the corresponding equilibrium constant for the above reactions and a detailed treatment of carbonate chemistry has been also presented by Stumm and Morgan (1996) and by Langmuir (1997).

Limestone is dissolved by water which contains dissolved CO_2 (or carbonic acid) and the reaction becomes:

$$CaCO_{3(s)} + H_2O + CO_{2(aq)} \leftrightarrow Ca^{2+}_{(aq)} + 2HCO^-_{3\,(aq)} \tag{4.7}$$

The reaction in Eq. 4.7 explains the formation of caves in calcareous terrains, like the extraordinary *cenotes* of the Mayan peninsula (Fig. 4.4). It is evident that if the equilibrium of Eq. 4.7 is displaced to the left-hand side of the chemical equation, due to Le Chatelier's principle, one mole of CO_2 is returned to the gaseous pool and another may be sequestered as calcite. At any rate, carbonation is a mechanism that exports alkalinity from the continents to the sea. Harmon and Wicks (2006) have brought together many case studies that deal with the geomorphologic, hydrologic, and geochemical aspects associated with carbonation processes.

4.2.4 Hydrolysis

Another significant mechanism in chemical weathering is the **hydrolysis** of minerals. It may occur under neutral, acidic, or basic conditions, with compounds in which either cations give rise to weak bases or anions give rise to weak acids, or both the cation and anion give rise to weak bases and acids. The suffix "-olysis" implies the chemical reaction or breaking down of a chemical compound, such as a silicate.

A typical acid hydrolysis, which is closely associated with the carbonation mechanism and which is of enormous significance in continental denudation is the *incongruent dissolution* of sodium feldspar to clay (i.e., kaolinite) by water saturated with CO_2:

$$\begin{aligned} &2NaAlSi_3O_{8(s)} + 9H_2O + 2H_2CO_{3(aq)} \leftrightarrow Al_2Si_2O_5(OH)_{4(s)} \\ &+ 2Na^+_{(aq)} + 2HCO^-_{3\,(aq)} + 4H_4SiO_{4(aq)} \end{aligned} \tag{4.8}$$

Gibbs free energy of formation values, which can be found in the literature (e.g., Faure 1998), allow to calculate $\Delta G^0 = 69.1$ kJ for the above reaction. The positive value of ΔG^0 means that the reaction is not energetically favored and, hence, it is not spontaneous. There is ample evidence, however, that it occurs. The corresponding equilibrium constant,

$$\log_{10}K = \Delta G^\circ/(2.303RT) = -12.1022 \tag{4.9}$$

R is a constant with the value of 8.314 $JK^{-1}mol^{-1}$ and T is the temperature in degrees Kelvin; 2.303 is the conversion factor to transform ln K into $\log_{10} K$.

Therefore,

$$K = 7.910^{-13} \tag{4.10}$$

The value of K shows that a relatively large quantity of carbonic acid is needed for the reaction to occur. Hence, the scenario that comes into focus is one of an aqueous solutions moving away from the mineral surface under chemical attack and a significant replenish rate of fresh carbonated water coming into contact with the mineral surface, thus sustaining the continuous hydrolysis process. This chemical mechanism plays a key role in the transfer of major dissolved ions via riverine or groundwater pathways to the world's oceans (Meybeck 2005).

Many authors have probed into the significance of silicate hydrolysis in the overall picture of continental chemical weathering, (Fig. 4.5) and there are many examples that can be cited (e.g., Zakharova et al. 2005; Banks and Frengstad 2006; Zakharova et al. 2007; Rajamani et al. 2009) as important contributions to increase the present knowledge. At any rate, White (2005) has completed a valuable in-depth review on the natural weathering of silicate minerals.

Chemical weathering not only supplies major chemical components to the hydrosphere, like cations and anions. They are accompanied by numerous trace elements (i.e., concentrations lower than 1 ppm), which usually also are trace

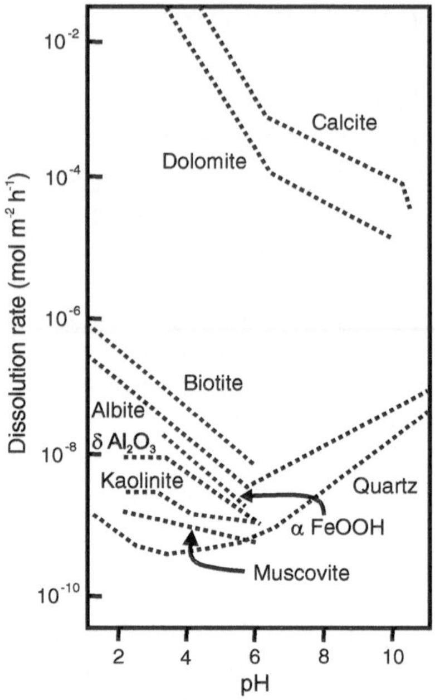

Fig. 4.5 Dissolution rates of different minerals as a function of pH (25 °C). For data, see Stumm and Morgan (1996). Reproduced with permission, John Wiley and Sons

elements in rocks, although there are some exceptions: Fe, Al, and Ti are major rock components but, due to their low solubility, occur as traces in natural waters (Gaillardet et al. 2005). Of special interest among trace components are the group or rare earths (REE), which are freed during weathering but are rapidly adsorbed onto colloids and thus constitute valuable tracers of the original rock (e.g., Elderfield et al. 1990; Sholkovitz 1995) (Fig. 4.6).

4.2.5 Oxidation-Reduction Processes

Redox reactions are common in aqueous solution, whether as weathering reactions or in the early stages of sediment diagenesis, in the water-mud interface (e.g., Potter et al. 2005). Although oxidation and reduction occur jointly, in weathering processes one product may be more relevant than others, so the following treatment of the subject matter is divided into oxidation (i.e., when oxidized products are the significant part) and reduction (i.e., when reduced phases are the more important byproducts). One interesting aspect of this weathering mechanism (it also happens in early *diagenetic processes*) is that it is often mediated by microbes (as seen in Chap. 3) and organic debris.

Oxidation usually occurs with free oxygen (i.e., dissolved in water) as the oxidizing agent. For example, in the reaction:

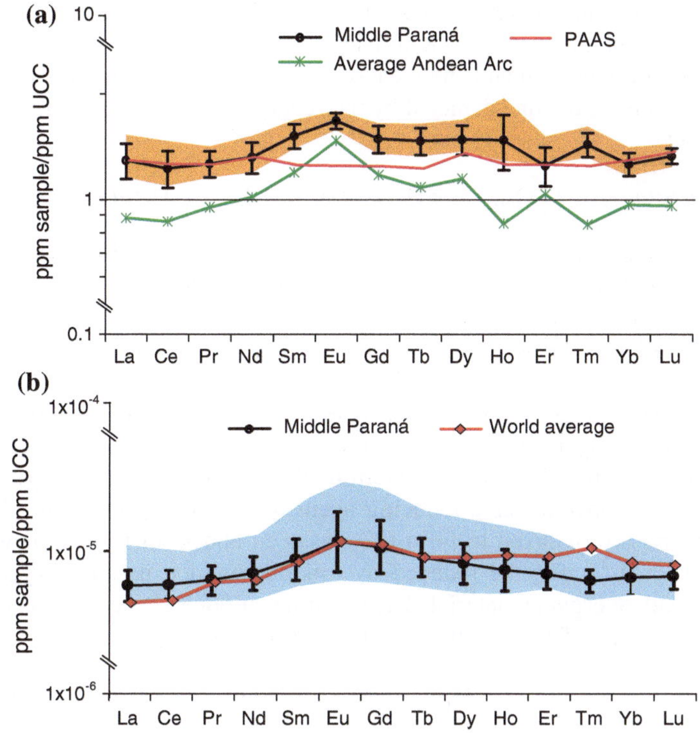

Fig. 4.6 Upper continental crust (UCC)-normalized REE spidergrams for Paraná River. **a** Paraná total suspended sediments (shaded area, N = 35, data from Depetris and Pasquini 2007); Post Archean Australian shale (PAAS, Taylor and McLennan 1985); and Average Andean Arc (AAA, http://www.geokem.com/) normalized compositions are included for comparison. Note the pronounced Eu positive anomaly in the Paraná River and in the AAA. **b** Dissolved REE in the middle Paraná River (data from Depetris and Pasquini 2007); world average dissolved REE composition (Gaillardet et al. 2005) is included for comparison. Note that the dissolved fraction preserves the Eu positive anomaly observed in total suspended sediments

$$4FeO_{(s)} + O_{2(g)} \rightarrow 2Fe_2O_{3(s)} \quad (4.11)$$

The corresponding value for $\Delta G^0 = -240$ kJ. The associated equilibrium constant for the reaction is $K = 1.08 \times 10^{42}$, thus indicating that the reaction will go practically to completion, as direct observation of rusting iron proves.

Sulfides are common in coal deposits, mineral veins, and mudrocks; for example, the oxidation of Fe(II) and S in pyrite generates sulfuric acid:

$$2FeS_{2(s)} + 7^1/_2O_{2(g)} + 7\ H_2O \rightarrow 2Fe(OH)_{3(s)} + 4H_2SO_{4(aq)} \quad (4.12)$$

As was presented in Chap. 3, microorganisms play a very significant role and are closely involved in sulfide oxidation. The resulting acidity (e.g., pH may be as low as 1 or 2; according with Norstrom et al. (2000) it may be even negative in natural environments) enhances the solubility of aluminum and other metals which causes toxicity in aquatic ecosystem. Clearly, this mechanism, when active, promotes mineral dissolution and chemical denudation. Another example of oxidation worth addressing is the one depicted by the reduced iron-bearing fayalite, the iron-rich olivine:

$$Fe_2SiO_{4(s)} + {}^1\!/_2O_{2(g)} + 5H_2O \rightarrow 2Fe(OH)_{3(s)} + H_4SiO_{4(aq)} \quad (4.13)$$

The products of this reaction are silicic acid and colloidal hydrated iron oxide which, after dehydration, yields a variety of iron oxides whose common occurrence reflects their insolubility under the oxidizing Earth surface conditions. Both phases—dissolved and solid—are common participants in global continental denudation.

In weathering processes, **reduction** is less significant than oxidation reactions. It occurs frequently in anaerobic environment-types, such as waterlogged soils whose greenish-grayish color is determined by reduced iron oxides. Organic matter typically operates as a reducing agent in weathering processes. Tissue or organic debris are oxidized to form CO_2, as seen above, or to form new organic compounds. $SO_4{}^{2-}$ may be reduced by microbes which use the oxygen in the sulfate to oxidize organic material. The resulting sulfide (S^{2-}) can go on to react in various ways. One of them is the reaction that may take place:

$$2H^+_{(aq)} + S^{2-}_{(aq)} \rightarrow H_2S_{(g)} \quad (4.14)$$

Besides entering the gas pool, sulfide can react with metals to precipitate, for example, FeS or MnS (Fig. 4.7).

In addition to their capability of alternately mobilizing or immobilizing metals associated with naturally occurring aquifer materials, redox processes can also

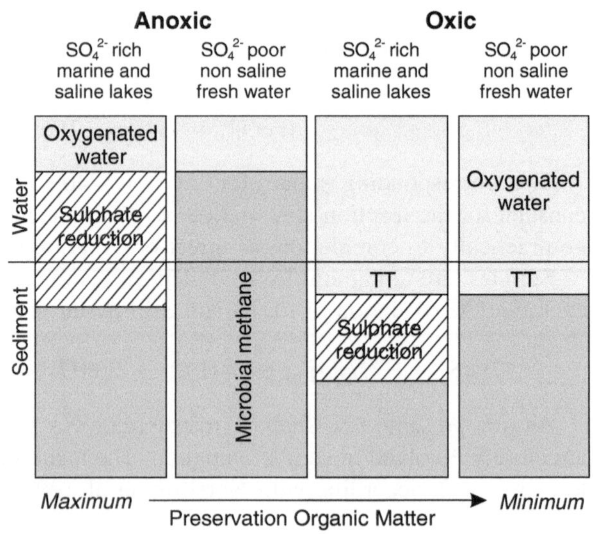

Fig. 4.7 A comparison between typical anoxic and oxic aqueous environments. T represents turbulence (Potter et al. 2005). Note the location of the sulfate reduction zone in both, anoxic and oxic environments; organic matter is better preserved in saline sulfate rich environments. Reproduced with permission from Springer Science

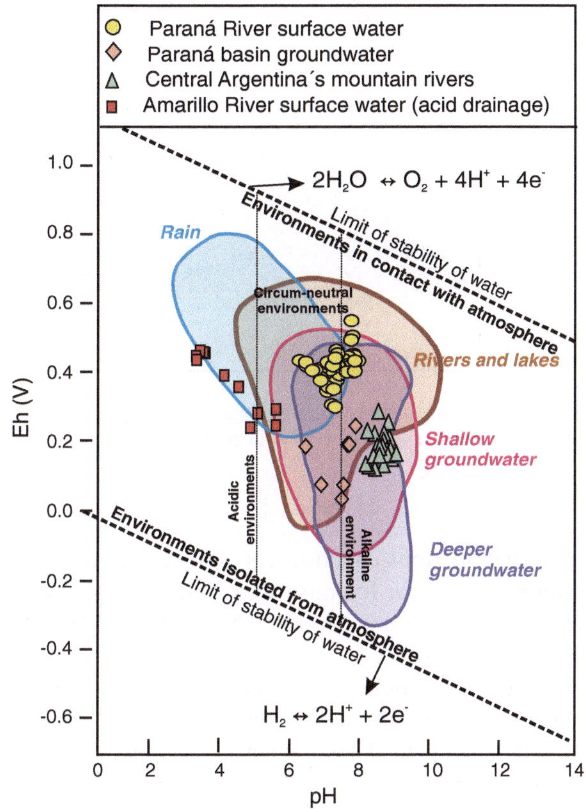

Fig. 4.8 pH-Eh (i.e., a measure of the redox state of a solution) diagram showing the fields for natural waters. Data from different environments have been added for comparison (Pasquini and Depetris 2012 and unpublished sources). The diagram shows the evolution of waters through the cycle as it becomes less oxic and more buffered with respect to pH

affect the chemical quality of ground water contributing to the generation of byproducts such as Fe^{2+}, H_2S, CH_4, and to the preservation or degradation of anthropogenic contaminants (e.g., McMahon and Chapelle 2008) (Fig. 4.8).

4.2.6 Exchangeable Ions

The fine-grained mineral particles that result from the weathering of minerals and rocks are particularly important in supplying a reservoir of exchangeable cations and anions, which may be involved in supplementary weathering processes. These clays (particle size <2 μm) and colloids (particle size <0.2 μm) may be mineral matter or humus and carry many negative charges on their surface, and individual positive exchangeable ions attached to the surface. **Cation exchange** (e.g., H^+, Na^+, Ca^{2+}, Al^{3+}, etc.) is more common than anion exchange (Fig. 4.9).

Applying the mass law to exchange reactions:

$$\{Na^+R^-\} + K^+ \leftrightarrow \{K^+R^-\} + Na^+ \quad (4.15)$$

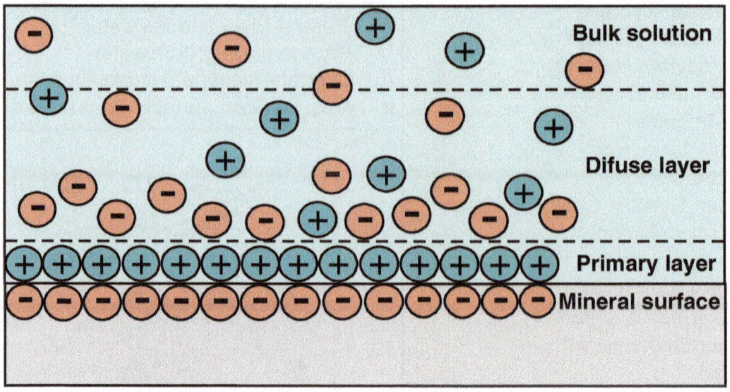

Fig. 4.9 Schematic representation of a diffuse-layer adsorption model showing the association of ions to an exchange surface such as a clay particle

$$2\{Na^{+}R^{-}\} + Ca^{2+} \leftrightarrow \{Ca^{2+}R_2^{-}\} + 2\ Na^{+} \tag{4.16}$$

In the previous equations, R^{-}symbolizes the negatively charged network of the cation exchanger; for most clays the general order of ion affinity (known as the Hofmeister series) is $Cs^{+}>K^{+}>Na^{+}>Li^{+}$, and $Ba^{2+}>Sr^{2+}>Ca^{2+}>Mg^{2+}$ (Stumm and Morgan 1996).

Different particles (or different soils) have different *cation exchange capacity* (or CEC), which is the concentration of cations in milliequivalents per 100 g of soil or sediment. These adsorbed cations may have an effect on the chemical composition of the soil solution, which in turn may modify weathering processes in the regolith.

Adsorption and desorption processes may be important controlling isotope fractionation in soil profiles, as was shown for Mg isotopes in an extreme weathering scenario, in the basaltic island of Hainan, China (Huang et al. 2012).

Glossary

Cation exchange capacity: Is the maximum quantity of total cations that a soil or sediment is capable of holding, and is available for exchange with the soil solution at a given pH value.

Cenote: Is a deep natural pit, or sinkhole, characteristic of Yucatán (Mexico) that results from the collapse of limestone, exposing the groundwater underneath.

Congruent dissolution: Weathering reaction between a mineral and water that results in its complete dissolution.

Diagenetic processes: Are changes that occur in sediment or sedimentary rocks during and after rock formation (lithification), at temperatures and pressures less than that required for the formation of metamorphic rocks or melting.

Hydrate: A substance that contains water of crystallization or water of hydration, in a definite ratio as an integral part of the crystal.

Incongruent dissolution: Weathering reaction between a mineral and water that results in its partial dissolution and a solid residue.

Karst: Is a geological formation shaped by the dissolution of soluble bedrock, usually carbonate rock but also in gypsum. Given the right conditions it can also occur in weathering-resistant rocks, such as quartzite.

Monomer: Is a molecule that may bind chemically to other molecules to form a polymer.

Playa: Is an ephemeral lakebed, or a remnant of an endorheic lake, consisting of fine-grained sediments infused with alkali salts. Alternative names include dry lake or alkali flat.

Regolith: I Layer of loose, heterogeneous material covering solid rock. It includes dust, soil, broken rock, and other related materials and is present on Earth, the Moon, Mars, some asteroids, and other terrestrial planets and moons

Solubility product: Is a constant (K_{sp}) for a solid substance dissolving in an aqueous solution; represents the level at which a solute dissolves in solution.

References

Alkattan M, Oelkers EH, Dandurand JL et al (1997) Experimental studies of halite dissolution kinetics, 1 the effect of saturation state and the presence of trace metals. Chem Geol 137(3–4):201–219

Anderson SP, von Blackenburg F, White AF (2007) Physical and chemical control on the critical zone. Elements 3:315–319

Banks D, Frengstad B (2006) Evolution of groundwater chemical composition by plagioclase hydrolysis in Norwegian anorthosites. Geochim Cosmochim Acta 70:1337–1355

Benedetti MF, Dia A, Riotte J et al (2003) Chemical weathering of basaltic lava flows undergoing extreme climatic conditions: the water geochemistry record. Chem Geol 201:1–17

Depetris PJ, Pasquini AI (2007) The geochemistry of the Paraná River: an overview. In: Iriondo MH, Paggi JC, Parma MJ (eds) The middle Paraná River: limnology of a subtropical wetland. Springer-Verlag, Berlin

Drever JI (1997) The geochemistry of natural waters. Surface and groundwater environments, 3rd edn. Prentice Hall, Upper Saddle River

Drever JI (2005) (ed) Surface and ground water, weathering, and soils. Elsevier, Amsterdam

Elderfield H, Upstill-Goddard R, Sholkovitz ER (1990) The rare earth elements in rivers, estuaries, and coastal seas and their significance to the composition of ocean waters. Geochim. Comsmochim Acta 54:971–991

Faure G (1998) Principles and applications of geochemistry, 2nd edn. Prentice Hall, Upper Saddle river

Forti Adolpho MC, Astolfo JMR, Fostier AH (2012) Rainfall chemistry composition in two ecosystems in the northeastern Brazilian Amazon (Amapá State), doi: 10.1029/2000JD900235

Gaillardet J, Dupré B, Allègre CJ (1999) Geochemistry of large river suspended sediments: Silicate weathering or recycling tracer? Geochim Cosmochim Acta 63(23/24):4037–4051
Gaillardet J, Viers J, Dupré B (2005) Trace elements in river waters. In: Drever JI (ed) Surface and ground water, weathering, and soils. Elsevier, Amsterdam
Gavrieli I, Starinsky A, Bein A (1989) The solubility of halite as a function of temperature in the highly saline Dead Sea brine system. Limnol Oceanogr 34(7):1224–1234
Harmon RS, Wicks CM (2006) Perspectives on karst geomorphology, hydrology, and geochemistry: a tribute volume to Derek C. Ford and William B. White. Geological society of America, Boulder
Huang KJ, Teng FZ, Wei GJ et al (2012) Adsorption- and desorption-controlled magnesium isotope fractionation during extreme weathering of basalt in Hainan Island. China. Earth Planet Sc Let. 359–360:73–83
Krauskopf KB (1956) Dissolution and precipitation of silica at low temperatures. Geochim Cosmochim Acta 10:1–26
Krauskopf KB, Bird DK (1995) Introduction to geochemistry, 3rd edn. McGraw-Hill, New York
Langmuir D (1997) Aqueous environmental chemistry. Prentice Hall, Upper Saddle River
Meybeck M (2005) Global occurrence of major elements in rivers. In: Drever JI (ed) Surface and ground water, weathering, and soils. Elsevier, Amsterdam
McMahon PB, Chapelle FH (2008) Redox processes and water quality of selected principal aquifer systems. Ground Water 46(2):259–271
Norstrom DK, Alpers CN, Ptacek CJ, et al. (2000) Negative pH and extremely acidic mine waters from Iron Mountain, California. Environ Sci Technol 34:254–258
Pasquini AI, Depetris PJ (2012) Hydrochemical considerations and heavy metal variability in the middle parana′ river. Environ Earth Sci 65:525–534
Piccini L, Mecchia M (2009) Solution weathering rate and origin of karst landforms and caves in the quartzite of Auyan-tepui (Gran Sabana, Venezuela). Geomorphology 106:15–25
Potter PE, Maynard JB, Depetris PJ (2005) Mud and mudstones. Introduction and overview. Springer, Berlin
Rajamani V, Tripathi JK, Malviya VP (2009) Weathering of lower crustal rocks in the kaveri river catchment, southern India: Implications to sediment geochemistry. Chem Geol 265:410–419
Sholkovitz ER (1995) The aquatic geochemistry of rare earth elements in rivers and estuaries. Aquat Chem 1:1–34
Stallard RF, Edmond JM (1987) Geochemistry of the amazon: 3. Weathering chemistry and limits to dissolved inputs. J Geophys Res 92:8293–8302
Stumm W, Morgan JJ (1996) Aquatic chemistry. Chemical equilibria and rates in natural waters, 3rd ed. Wiley-Interscience, New York
Taylor SR, McLennan SM (1985) The continental crust: its composition and evolution. Blackwell, Oxford
White AF (2005) Natural weathering rates of silicate minerals In: Drever JI (ed) Surface and ground water, weathering, and soils, Elsevier, Amsterdam
White AF, Brantley SL (ed) (1995) Chemical weathering rates of silicate minerals. Reviews in Mineralogy, Vol. 31. Mineralogical Soc. of America, Washington DC
Wray RAL (1997) A global review of solutional weathering forms on quartz sandstones. Earth-Science Rev. 42:137–160
Zakharova EA, Pokrovsky OS, Dupré B et al (2007) Chemical weathering of silicate rocks in Karelia region and Kola Peninsula, NW Russia: Assessing the effect of rock composition, wetlands and vegetation. Chem Geol 242:255–277
Zakharova EA, Pokrovsky OS, Dupré B et al (2005) Chemical weathering of silicate rocks in aldan shield and baikal uplift: insights from long-term seasonal measurements of solute fluxes in rivers. Chem Geol 214:223–248
Zanor GA, Piovano EL, Ariztegui D et al (2013) El registro sedimentario Pleistoceno Tardío-Holoceno de la Salina de Ambargasta (Argentina central): una aproximación paleolimnológica. Revista Mexicana de Ciencias Geológicas 30(2):336–354

Chapter 5
Weathering: Intensity and Rate

Abstract The evaluation of the intensity of weathering is usually achieved by means of numerous procedures, which may be absolute or relative; may be tackled by examining the solid residue left by weathering or by establishing the nature of the dissolved fraction, whose largest proportion is exported from the continents via streams, rivers, and ground waters. Absolute methods are feasible when weathering profiles are complete and clearly exposed whereas relative methodologies are useful when the weathered product is not necessarily near its source. The use of multivariate methodologies and modeling appears as promising techniques to assess weathering intensity. Laboratory experimentation, on the other hand, has supplied useful information on weathering rates of minerals. The field-supported mass balance approach, however, has furnished the most reliable information on weathering rates.

Keywords Weathering indexes · Laterite · Mineral stability · weathering sequence · CIA · WIP · ICV · Alpha indexes · PHREEQC · Chemical modeling

5.1 Introduction

The intensity or degree of weathering is the alteration, added together, attributable to diverse intervening processes that modify significantly the original solid rock or sediment, producing solid—usually modified—and dissolved products. Basically, weathering intensity refers to the degree of decomposition of minerals and rocks at certain point in time. In contrast, rate alludes to the amount of change (e.g. of mass) per unit time. Obtaining a rate (e.g., d*m*/d*t*, where *m* is mass and *t* is time), requires knowledge of the time period during which alteration has taken place and, therefore, this is a difficult task in the Earth Sciences, although it may be approachable in the laboratory. Intensity and rate are associated when a high weathering intensity may involve a relatively fast alteration rate. However, this is not always the case because the humid tropics, for example, exhibit a high

P. J. Depetris et al., *Weathering and the Riverine Denudation of Continents*, SpringerBriefs in Earth System Sciences, DOI: 10.1007/978-94-007-7717-0_5,

weathering intensity, sometimes achieved by moderate rates acting over an extended time period.

In this chapter, we will examine the best known methodologies available to determine quantitatively weathering intensity and rates.

5.2 Weathering Intensity

The concept of weathering intensity involves the degree of alteration that a rock or sediment exhibits at a certain point in time, after being exposed to weathering agents. When the intensity of alteration appears to remain stable for an extended time period, reflecting a balance between external processes and material properties, the system is said to be in equilibrium. In weathering, true equilibrium is seldom maintained, mainly because external processes are often dynamic and change over time. However, in tectonically stable parts of the humid tropics, with prolonged periods of relatively undisturbed weathering, a sort of equilibrium is achieved, with thick *lateritic* profiles rich in Al_2O_3 and Fe_2O_3 (Fig. 5.1). The dominant mineral species are quartz, goethite [FeO(OH)], gibbsite [$Al(OH)_3$], or kaolinite. Gibbsite and quartz do not coexist and kaolinite forms whenever silica is available. Thus, the general equation (Potter et al. 2005) that synthesizes the overall process of weathering, from initiation to final stages is:

$$\begin{aligned} H^+ + \text{primary mineral} &\rightarrow \text{intermediate clay mineral} + \text{solution} \\ &\rightarrow \text{gibbsite} + \text{solutions} \end{aligned} \quad (5.1)$$

The question of weathering intensity becomes complex when one considers that different minerals respond differently to weathering. This observation led to the idea that minerals can be arranged in order of their persistence through time, a condition, which may be inferred as a proxy for stability. Goldich (1938) derived, mainly from field observations, a mineral stability series for common rock-forming minerals that is still used as a general guide (Fig. 5.2). This sequence of progressively more stable minerals is analogous to Bowen's sequence of mineral crystallization from a melt. This is associated with the notion that the stability of a mineral species increases as the difference between the temperature of formation and that prevailing in the weathering environment decreases. Some studies, however, have shown that, depending on the environmental conditions, there are discernible deviations from Goldich's sequence.

A few years after Goldich's paper was published, Pettijohn (1941) attempted a sequence of mineral persistence based on mineral frequency in sedimentary rocks of increasing age, abandoning the two-tiered approach used by Goldich. In spite of the above-mentioned limited approval, Goldich's sequence has proved to be more accepted than Pettijohn's.

It is clear, then, that a factor like climate, which embraces or is connected with other aspects, like ambient temperature, water availability and chemistry, and

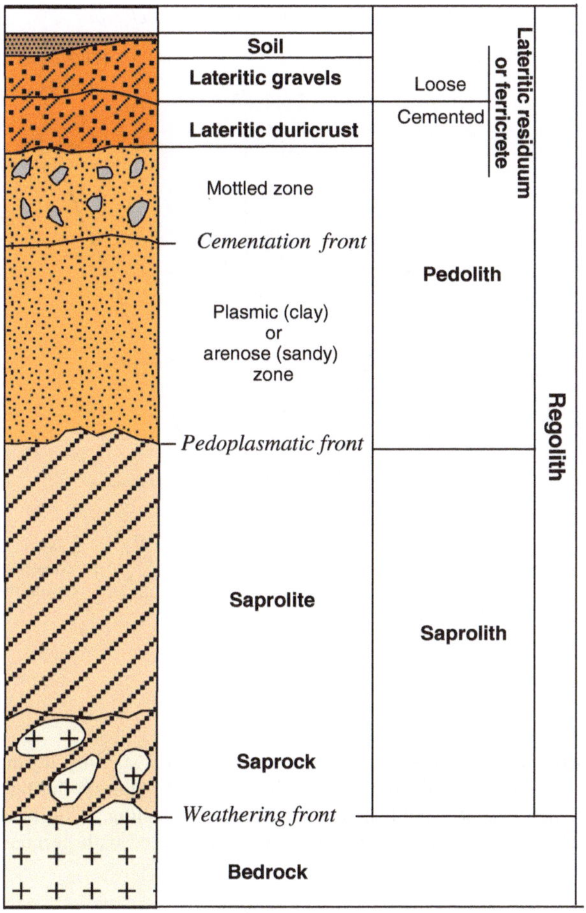

Fig. 5.1 Idealized weathering profile evolving toward a mature laterite. (Modified from McQueen and Scott 2009). © CSIRO 2008. Published by CSIRO Publishing, Collingwood, Victoria, Australia, http://www.publish.csiro.au/pid/5955.htm. Reproduced with permission

geomorphology, turns out to be very important at the time of assessing weathering intensity. The dynamics steering weathering, especially the temperature factor, the chemical nature of reacting water, and hydrodynamics vary globally and the expectation is that wide alteration zones would develop as an outcome (Fig. 5.3).

The assessment of weathering intensity has used methodologies that vary from the purely descriptive approach (e.g., fresh rock, slightly weathered, moderately weathered, etc.) to the objectivity that brings about scientific methodology.

5.2.1 *Measures of Weathering Intensity Based on Absolute Methods*

As weathering progresses and the more mobile elements are leached, the chemical composition of the remaining regolith will evolve. To evaluate the stage of such

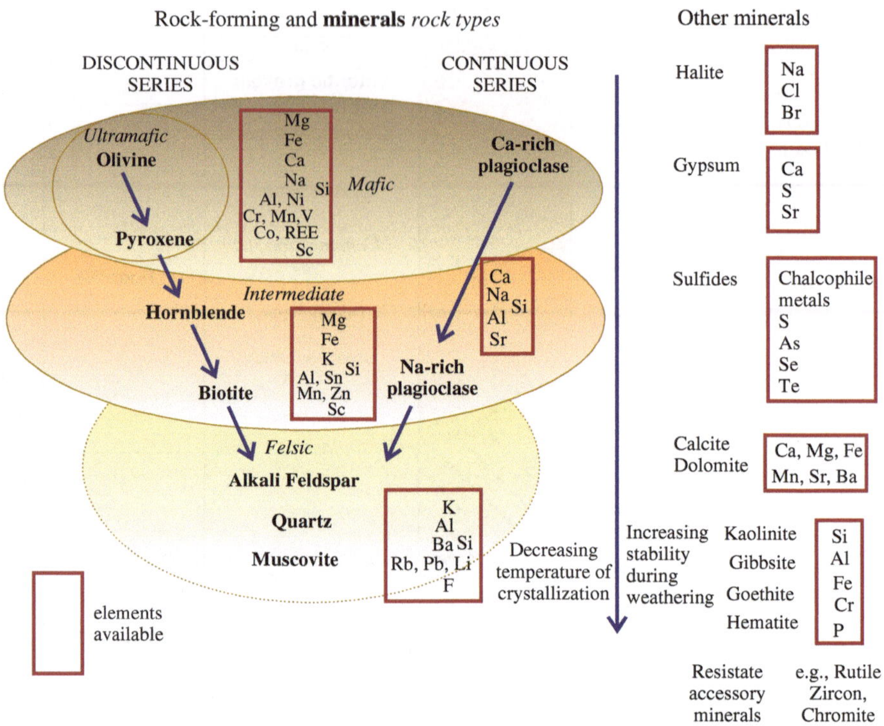

Fig. 5.2 Bowen's reaction series and mineral stability during weathering. Decreasing temperature of crystallization results in increased stability during weathering (McQueen 2009). © CSIRO 2008. Published by CSIRO Publishing, Collingwood, Victoria, Australia, http://www.publish.csiro.au/pid/5955.htm. Reproduced with permission

evolution, there is one approach that compares the constitution of the parental rock with that of the weathered material; these would be the "absolute" methods.

The isovolumetric method is based on the assumption that there has been no change in the sequence unweathered (or fresh) → weathered (or altered) states. The assumption should be supported by the preservation of the original petrographic textures or geological structures in the altered material. It follows that this would be the case when weathering intensity is very low, as it was the situation with coarse granitic *saprolites* of Portugal, when it was first applied (Braga et al. 1990). The method involves a comparison of the content of the main oxides between the fresh and weathered samples; the difference (i.e., a loss of chemical elements in the altered sample) is a measure of the intensity of weathering.

The other "absolute" method is the so-called "benchmark" method and it makes use of the insolubility of Al_2O_3 (the "benchmark") over a wide pH range as a means to calculate the relative losses (or gains) of the remaining oxides. This is accomplished using the ratio between the alumina content of the parent rock and of

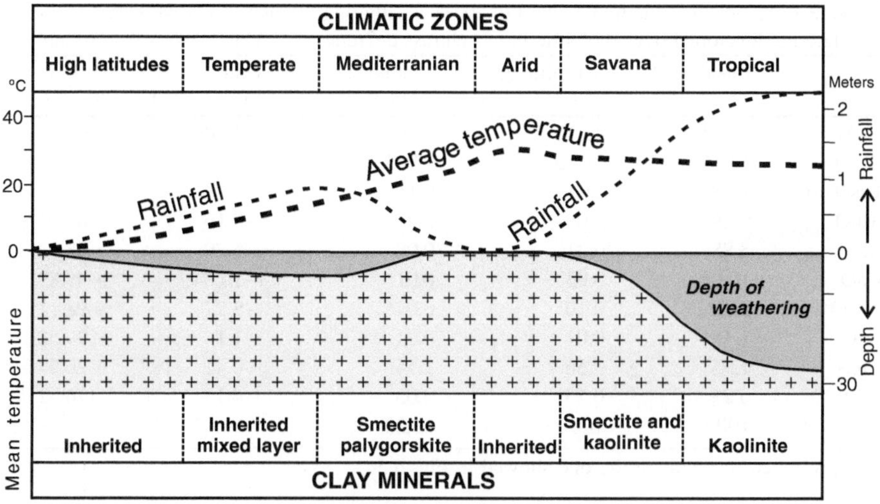

Fig. 5.3 Schematic representation of N–S depth of weathering and main associated clay minerals. Note the role of rainfall and temperature (Potter et al. 2005). Reproduced with permission from Elsevier and Springer Science

the regolith to calculate the relative change occurred in the remaining oxides (e.g., Krauskopf and Bird 1995).

Table 5.1 shows a model exercise performed in the tholeiitic basalt exposed to the tropical climate of Misiones (Argentina) that shows in the laterite a relative loss of ~57 % as soluble material, which is close to similar calculation performed by Faure (1998) in the same rocks (Basalts from Paraná, Brazil).

5.2.2 *Measures of Weathering Intensity Based on Relative Methods*

When there is uncertainty on the composition of the parent material or it is simply unknown, the level of weathering can be assessed by calculating the ratio between the more stable and the less stable minerals. These are known as **relative methods** and can be approached with the use of resistant minerals (e.g., tourmaline and zircon) for heavy (*h*) and quartz for light (*l*) minerals. The corresponding weathering ratios would be:

$$\mathrm{WR}_{(h)} = (\text{zircon} + \text{tourmaline})/(\text{pyroxenes} + \text{amphiboles}) \tag{5.2}$$

$$\mathrm{WR}_{(l)} = \text{quartz}/\text{feldspars} \tag{5.3}$$

Most relative approaches, however, have resorted to chemical methodology; based on the assumption that alumina remains immobile during weathering, a

Table 5.1 "Absolute" method applied to assess weathering intensity (relative losses and gains) in a laterite developed over tholeiitic basalt (Jurassic-Cretaceous) from Misiones (Argentina)

	Basalt (%)	Saprolite (%)	Amount remaining (g)	Gain (+) or loss (−) (g)	Gain (+) or loss (−) (%)
SiO_2	50.71	43.17	18.66	−32.05	−63.20
Al_2O_3	13.63	31.53	13.63	0.00	0.00
Fe_2O_3 (t)	13.89	21.17	9.15	−4.74	−34.12
MnO	0.19	0.08	0.03	−0.16	−82.01
MgO	5.88	0.21	0.09	−5.79	−98.42
CaO	10.05	0.03	0.01	−10.04	−99.89
Na_2O	2.32	0.10	0.04	−2.27	−98.11
K_2O	1.15	0.01	0.01	−1.14	−99.52
TiO_2	1.95	3.56	1.54	−0.41	−20.94
P_2O_5	0.23	0.13	0.05	−0.18	−76.41
Sum	100	100	43.23	−56.77	

The profile corresponds to the one showed in Fig. 5.5

number of indices have been proposed, from simple ratios to more complex equations.

Chittleborough (1991) developed an interesting overview of the most common indices used until that time. The Ruxton ratio (Ruxton 1968), for example, is the simplest approach, involving the molar ratio of silica to alumina:

$$R = SiO_2/Al_2O_3 \tag{5.4}$$

This ratio relates the loss of SiO_2 to total-element loss and seems to work nicely in weathering profiles developed on acid to intermediate rocks in humid regions, where the final products are kaolin and/or allophane.

Vogt's ratio or residual index (Vogt 1927, cited by Price and Velbel 2003) uses a more elaborate approach, comparing relatively immobile oxides with the more soluble phases:

$$V = (Al_2O_3 + K_2O)/(CaO + MgO + Na_2O) \tag{5.5}$$

Price and Velbel (2003) cite Roaldset's (1972) work, who used this index to compare the bulk chemistry of moraine clays with marine clays in Quaternary deposits in Norway.

The modified weathering potential index (MWPI) included more oxides in the calculation:

$$MWPI = [(CaO + MgO + Na_2O + K_2O)/(CaO + MgO + Na_2O + K_2O + SiO_2 + Al_2O_3 + Fe_2O_3)]100 \tag{5.6}$$

Indices that include iron in their calculations, however, must be used with great care because they do not distinguish between Fe(II) and Fe(III) (i.e., all iron is expressed as Fe_2O_3), and the oxidation state and distribution of iron in a weathering profile is biased by the relative abundance of each oxidation state in, for example, a preweathered metamorphic rock. The interaction with ground water during weathering is also important in defining the oxidation state of iron.

The weathering index of Parker or WIP (e.g., Hamdan and Burnham 1996) is based on the proportions of alkali and alkaline earth metals:

$$\mathrm{WIP} = 100[(2\mathrm{Na_2O}/0.35) + (\mathrm{MgO}/0.9) + (2\mathrm{K_2O}/0.25) + (\mathrm{CaO}/0.7)] \tag{5.7}$$

Values of WIP are commonly between ~100 and 0 with the least weathered rocks having the highest values. It implicitly presupposes as well that all calcium in a silicate rock is contained in silicate minerals; this assumption is not a problem as long as there are no carbonate detritus or cements in the analyzed sample. Parker (1970) considered the susceptibility to weathering of the elements involved in the equation by including the denominator values of bond strength, as a measure of the energy necessary to break the cation-to-oxygen bonds of the respective oxides. These coefficients are considered to reflect the probability of an element to be mobilized during the weathering process.

This index was advantageously employed to evaluate, along with the chemical index of alteration (CIA), the chemical leaching in suspended particulate matter of Chinese rivers (Shao et al. 2012). Strømsøe and Paasche (2011) also used the CIA and the WIP in high-latitude regolith and inferred that chemical weathering gradually controls the production of fine silt, very fine silt, and clay, whereas physical weathering primarily controls the production of grain size fractions larger than 32 μm, a condition that seems to be an intrinsic feature in the formation of weathered high-latitude regolith.

The chemical index of weathering (CIW) (Harnois 1988) is another relative "benchmark" method that has been used to assess the intensity of weathering. The value of this index increases as the degree of weathering increases, and the difference between CIW values of the silicate parent rock and soil or sediment reflects the amount of weathering experienced by the weathered material:

$$\mathrm{CIW} = [\mathrm{Al_2O_3}/(\mathrm{CaO} + \mathrm{Na_2O} + \mathrm{Al_2O_3})]100 \tag{5.8}$$

This method is still in use and has proved to be a valuable approach in many studies. The work of Buggle et al. (2011) is an interesting example on the use of CIW and other weathering indices in loess-paleosols studies concluding that, due to diagenetic effects (*illitization*) most indices involve uncertainties. CIW (a.k.a. chemical proxy of alteration or CPA), however, is proposed as the more appropriate geochemical alternative to assess silicate weathering in the so-called loess-paleosols sequences. Another example involving soils is presented by Ekosse and Ngole (2012) who advantageously used the CIW (along with the chemical index of alteration or CIA): high values of the indices and low contents of alkali and alkaline earth elements suggested low amounts of essential constituents present in geographic soils of Swaziland. In a final example for this index, Pola et al. (2012) used CIW in the assessment of the influence of alteration on physical properties of volcanic rocks. They employed the index with success to interpret chemical heterogeneity, which is correlated with petrographical, physical, mineralogical, and mechanical information.

Plagioclase is abundant in silicate rocks and its Ca-rich end-member hydrolyzes relatively rapidly. Hence, the plagioclase index of alteration or PIA was proposed as an option to the CIW (Fedo et al. 1995):

$$PIA = 100[(Al_2O_3 - K_2O)/(CaO + Na_2O + Al_2O_3 - K_2O)] \quad (5.9)$$

The chemical index of alteration or CIA is probably the most accepted index to evaluate weathering intensity (Nesbitt and Young 1982):

$$CIA = [Al_2O_3/(CaO^* + Na_2O + K_2O + Al_2O_3)]100 \quad (5.10)$$

CaO^* represents CaO adjusted for apatite and Ca-bearing carbonates using P_2O_5 and CO_2 in the correction procedure (Fedo et al. 1995). CIA values of 45 to 55 indicate practically no weathering (the average upper crust has a CIA of 47). Kaolinite, gibbsite, chlorite, and bohemite have CIA values of $\sim$100, whereas smectite and illite groups have values ranging from about 70 to 80. Primary minerals have much lower CIA values: plagioclase and K-feldspars are 50; biotite, 50 to 55; amphibole and pyroxene, 0 to 20; and basaltic to rhyolitic glass range from 35 to 55. It follows that the CIA value of a bulk sample will vary substantially depending on the proportions of clay minerals and primary minerals in it (e.g., Nesbitt and Markovics 1997). CIA has been extensively applied to gain insight in a number of geological problems, like to probe into, or to assess climate transitions. Goldberg and Humayun (2010), for example, concluded that, with appropriate care, the CIA is a useful tool for the appraisal of humidity conditions in the rock record. Similarly, Bahlburg and Dobrzinski (2011) used the CIA as a climatic proxy in the study of Neoproterozoic glacial deposits.

The CIA in paleoenvironmental studies has proved valuable. González-Alvarez and Kerrich (2012) made an interesting comparison between weathering intensity in the Mesoproterozoic (Belt-Purcell Supergroup, from Canada and the USA) and modern large-river systems, concluding that CIA values reflect a more aggressive chemical weathering since Proterozoic rivers had less sediment residence times due to the absence of a vegetation cover and, therefore, faster transport times than their modern counterparts. Another interesting example on the use of the CIA as a tool in paleoenvironmental studies was supplied by Shen et al. (2013) which used the CIA, along with additional methodology such as REE, to assess volcanism in South China during the Late Permian and its relationship to marine ecosystems and environmental change.

Although not free from controversy (Li and Yang 2010), the aspect of Earth Sciences where the CIA has been extensively used is its application to continental weathering and denudation (McLennan 1993), studying the variability of the CIA determined for the mean suspended solids load of large world rivers, assuming that the index is a proxy for the weathering factors and processes that occur on their respective drainage basins. The CIA, for example, was used in South America's

Paraná River (Depetris et al. 2003), in Patagonia's Chubut River and also coupled to other weathering indices (Pasquini et al. 2005; Depetris and Pasquini 2007), like the index of compositional variability or ICV (Cox et al. 1995):

$$ICV = (CaO + MgO + Na_2O + K_2O + Fe_2O_3 + MnO + Ti_2O)/Al_2O_3 \quad (5.11)$$

This index measures the abundance of alumina relative to the other major cations in rocks or minerals; silica is excluded to eliminate problems with quartz dilution. It has been primarily formulated to investigate secular changes in mud-rock composition on a single continental crustal block which reflects both, sedimentary recycling processes and changes through time in the composition of crystalline material being added to the sedimentary system.

Price and Velbel (2003) evaluated several weathering indices that basically employ alkali, alkaline earth metals, and alumina to comparatively assess their performance in evaluating the weathering of heterogeneous metamorphic rocks. They concluded that the weathering index of Parker (WIP) is the most appropriate index of alteration for application to felsic heterogeneous weathered regoliths, mainly because it does not assume the immobility of alumina, a drawback that Price and Velbel (2003) considered important at the time of judging weathering intensity.

The problem of measuring weathering intensity was dealt with by Gaillardet et al. (1999) by defining a separate index for each mobile element whose concentration in the sediment was compared with that of an immobile element whose magmatic compatibility is close to that of the mobile element (Hofmann 1988). For example, ratios of elements with similar magmatic compatibilities, such as Sm/Na, are expected to be less variable than ratios of chemical elements whose magmatic compatibilities are divergent (e.g., Th/Na). Gaillardet et al. (1999) characterized indices for all the mobile elements, similar to the following examples for Na and Ca:

$$\alpha_{Na} = (Sm/Na)_{sed}/(Sm/Na)_{UCC} \quad (5.12)$$

$$\alpha_{Ca} = (Ti/Ca)_{sed}/(Ti/Ca)_{UCC} \quad (5.13)$$

A value of $\alpha_m \approx 1$ (where m is a mobile element) means that there is no net weathering and that the material (e.g., river-borne sediment) has a composition similar to the upper continental crust (UCC). In contrast, $\alpha_m > 1$ corresponds to depletion with respect to the UCC, and $\alpha_m < 1$ to enrichment of element m. (Fig. 5.4).

Using this approach, Gaillardet et al. (1999) found that α_{Na}, α_{Ba}, and α_K showed a global coherence and also that the highest weathering intensities are observed in the suspended solids of Orinoco, Niger, Seine, and Xijiang rivers. Less-weathered suspended sediments were found in rivers draining volcanic islands and in the Danube, Lena, Connecticut, Huanghe, Khatanga, and Yana. Sediment yield-weighted world mean weathering indices are $\alpha_{Na} = 7$, $\alpha_K = 1.7$, and $\alpha_{Ba} = 0.77$. (Fig. 5.5).

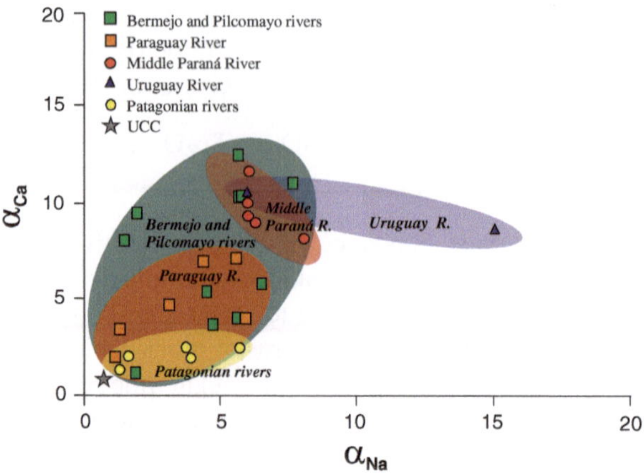

Fig. 5.4 Weathering α indices for calcium and sodium determined in the suspended load of different South American rivers (most data from Depetris and Pasquini 2007)

Multivariate statistical techniques [*principal component analysis* (PCA)] also have been used as a promising approach to assess the degree of weathering. The method authors (Ohta and Harai 2007) claimed that the *W* index has a number of noteworthy properties that are not found in other weathering indices. The *W* index appears to be receptive to weathering-induced chemical changes because it is based on eight major oxides, whereas most indices are defined by a smaller number (between two and four oxides). The authors also found that the *W* index provided robust results even for highly weathered samples, rich in *sesquioxides* (e.g., Al_2O_3, Fe_2O_3). Third, the *W* index seems appropriate for a wide range of *felsic*, intermediate, and *mafic* igneous rock types. Finally, the authors argue that the resulting diagrams facilitate provenance analysis of sedimentary rocks by recognizing their weathering trends and thereby allowing a backward estimate of the composition of the original unaltered source rock.

The assessment of weathering and denudation can also be approached from the point of view of dissolved phases. Tardy (1971), for example, calculated the

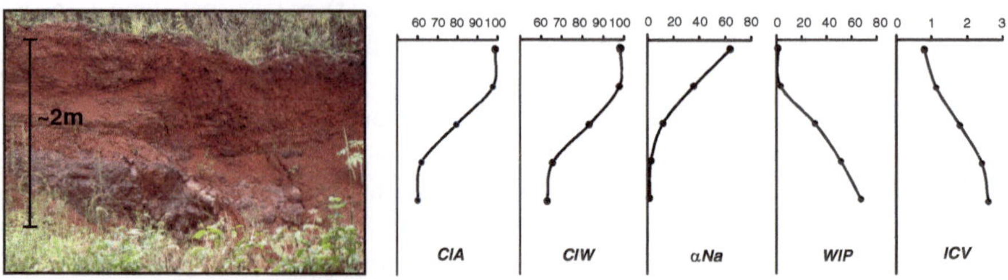

Fig. 5.5 Mature lateritic profile in Misiones (Argentina) developed over tholeiitic basalt (Serra Geral Formation). Associated curves show the evolution of different weathering indices throughout the soil profile. Photograph by P. J. Depetris

molecular ratio Re, using the concentrations of different dissolved elements measured in surface waters that drain granitic and gneissic catchments:

$$\mathrm{Re} = 3(\mathrm{Na}^+) + 3(\mathrm{K}^+) + 2(\mathrm{Ca}^{2+}) - (\mathrm{SiO}_2)/0.5(\mathrm{Na}^+) + 0.5(\mathrm{K}^+) + (\mathrm{Ca}^{2+}) \tag{5.14}$$

The coefficients used in Eq. 5.14 depend on the major primary minerals of the country rock and correspond to an average granitic composition with micas and feldspar. If Re $\approx$ 0, the dominant weathering product is gibbsite; if Re $\approx$ 2 the formation of kaolinite prevails; and if Re $\approx$ 4 smectites are the foremost weathering product. Boeglin and Probst (1998) adapted Eq. 5.14 and used it in the upper Niger River drainage basin to assess CO_2 consumption in a tropical lateritic environment. Their main conclusion was that the CO_2 flux consumed by silicate weathering was about two times lower in lateritic than in non-lateritic areas.

The complexities of geochemical modeling are beyond the scope of this monograph but it can be mentioned here that some computer programs, like PHREEQC (Parkhurst and Appelo 2013), are designed to perform a wide variety of aqueous geochemical calculations implementing several types of aqueous models. Using any of such aqueous models, PHREEQC has capabilities for (a) speciation and saturation-index calculations; (b) batch-reaction and one-dimensional transport calculations with reversible and irreversible reactions and specified mole transfers of reactants, kinetically controlled reactions, mixing of solutions, and pressure and temperature changes; and (c) inverse modeling, which finds sets of mineral and gas mole transfers that account for differences in composition between waters within specified compositional uncertainty limits.

Hundreds of scholarly articles quote or use of PHREEQC to evaluate geochemical and environmental problems in aquifers. Schettler et al. (2013), for example, used modeling as an aid to understand the geochemical evolution of a confined aquifer in the Amu Darya Delta (Aral Sea basin). Fewer articles use the computer code to assess weathering-controlled chemical losses in streams and rivers trough inverse modeling. A relevant case study was supplied by Lecomte et al. (2005) in Los Reartes River, a 250 km^2 mountainous drainage basin in a granitic and gneissic terrain, placed in the Comechingones range, Córdoba (Argentina). The study showed, among other conclusions, that oligoclase was the major solute supplier and kaolinite was the main precipitated phase in the granite domain, whereas muscovite was the chief solute supplier and illite was the precipitated phase in the gneissic realm. The steeper portions of the metamorphic reach are less crucial in supplying solutes than the lower ones, thus highlighting the importance of the water residence time in the kinetics of dissolution. CO_2 accounts for over 50 % of all species involved in the weathering reactions occurring in the studied basin. (Fig. 5.6) This is concurrent with results recently obtained in streams and rivers in the United States (Butman and Raymond 2011).

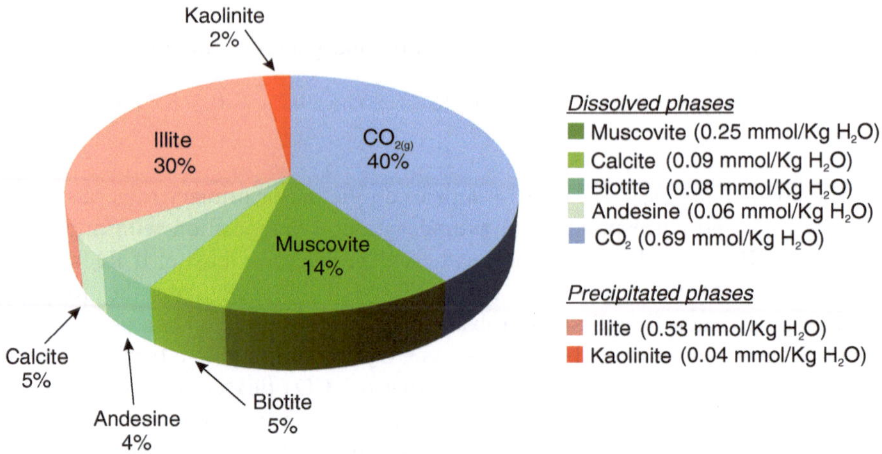

Fig. 5.6 PHREEQC inverse modeling exercise applied to Los Reartes River basin water chemistry data, (Córdoba, Argentina). The graph shows the percentages of dissolved and precipitated phases (minerals and CO_2), which explains the downstream water chemical evolution (data from Lecomte et al. 2005)

5.3 Weathering Rate

The assessment of weathering intensity is an aspect that is coupled with the rate at which minerals and rocks are disaggregated and dissolved when exposed to weathering agents. One way of undertaking this problem is through the **laboratory approach**, which usually has the difficulty that the rate of weathering reactions are slow and the solutions that must be employed are usually at much higher concentrations than in the environment. In any case, the weathering rate R (mol $m^{-2}s^{-1}$) of a primary silicate mineral is defined by the equation:

$$R = \Delta M / S t \tag{5.15}$$

ΔM (mol) is the mass change attributable to weathering, S (m^2) is the surface area involved, and t is the duration of the reaction. White (2005) has dealt with extensively with the determination and quantification of the terms on the right side of Eq. 5.15.

Goldich's weathering sequence (Fig. 5.2) is consistent with laboratory findings (e.g., Veibel 1993). From both approaches, it is clear that Ca-plagioclases weather faster than Na-plagioclases; that olivine weathers more rapidly than Ca-plagioclase; that amphiboles and Na-plagioclases weather at about the same rate; and that garnet weathers faster than Na-plagioclase (Fig. 5.7). Mass per unit time is typically used to describe rates, namely kg $km^{-2}yr^{-1}$; mol $g^{-1}h^{-1}$; meq $m^{-2}yr^{-1}$. For example, using published data obtained through experimental studies, a labradorite would weather at a rate of 4.9 10^{-7} mol $cm^{-2}yr^{1}$; the weathering of alkali feldspars would be one order of magnitude slower, 2.3 10^{-8} mol $cm^{-2}yr^{-1}$, and a

Fig. 5.7 Basilica of Santa Croce in Florence (Italy). The construction was completed in 1385. Several marble types were used in the façade: the white marble was quarried in Seravezza, and the dark serpentine marble from Pisa; the remaining marble types were quarried in other parts of Italy or imported from abroad. The inset shows the higher degree of weathering in the dark serpentine marble when compared with the white variety (Photograph by P. J. Depetris). The Basilica's image form Wikimedia Commons

hornblende would weather even slower, at a rate of 5.7 10^{-9}mol $cm^{-2}yr^{-1}$ (Velbel 1993). A linear measure of space is also used to describe weathering rates; granites appear to reach a rate of 6,000 to 13,000 μm 10^{-3}yr in central Europe, of 13,500 μm 10^{-3}yr in Chad (Africa), and of 5,000 to 50,000 μm 10^{-3}yr in Ivory Coast (Africa) (Colman and Dethier 1986).

To probe into naturally weathered materials is the other option, also looking into the chemistry of waters draining catchment areas. The difficulty inherent to this approach is that there are so many possible combinations of environmental conditions and reacting materials and, above all, the enormous difficulty associated with the isolation of the intervening factors.

Several approaches have been used in attempts to work out **weathering rates with field support**. One pathway is to study the degree of alteration in weathered profiles. In a regolith that is not undergoing significant erosion, a reaction front moves downward with time at a specific rate. The profile becomes quasi-stationary when the normalized concentrations (elemental or mineralogical) at the top of the profile no longer change while the profile may be moving downward at a certain weathering advance rate. If steady-state conditions prevail, the erosion rate at the top of the profile is equal to the weathering advance rate. This is the situation which is usually assumed for forested land, where most estimates for these rates vary between

7 and 80 mm per 100 yr, but extreme conditions may raise these estimates from one order of magnitude higher (Brantley 2005; Brantley et al. 2007).

The weathering rate in a lateritic profile, such as the one studied at the Niger drainage basin (Boeglin and Probst 1998) is in contrast with the situation described above, because the physical erosion rate at the top of the profile is lost at a rate of 2.4 to 3.4 m 10^{-6} yr^{-1}, whereas the weathering front at the bottom of the profile was estimated to move downward at a rate 4.7 m/10^{-6} yr^{-1}, thus indicating a net thickness increase for the soil profile. In the most humid an elevated areas of the Niger drainage basin, the laterite thickness appears to be in a steady state or even decreasing.

U-series isotopes were used in Puerto Rico to model the dynamics of a weathering front in volcaniclastic bedrock (Dosseto et al. 2012). The results showed that it takes 40–60 10^{-3} yr to develop a 18 m-thick profile, with a regolith production rate of 334 $\pm$ 46 mm 10^{-3} yr. When these rates are compared with weathering profiles developed over shale or granitic lithologies, it turns out that they are slower by a factor of 30.

Brantley et al. (2007) quoting other authors, reproduce regolith formation rates in areas with different lithology that fluctuate between 0.2–0.7 mm 10^{-2} yr and 5.8 mm 10^{-2} yr, with a world average of 7 mm 10^{-2} yr or 70 m 10^{-6} yr^{-1}. Amundson (2005) has developed a detailed treatment of soil formation, including intervening factors, soil morphology, mass balance models, matter and energy transfer, etc.

Using a system of geochemical mass balance equations assembled and constrained by petrological, mineralogical, hydrological, botanical, and aqueous geochemical data in natural forested watersheds has proved to be a successful approach to approximate natural rates. A valid example is the work of Velbel (1985) who found weathering front rates of the order of 3.8 cm 10^{-3} yr^{-1} in forested catchments in the Southern Blue Ridge (North Carolina, USA), and which agrees well with the "average" denudation rate of 4 cm 10^{-3} yr^{-1} determined by varied methodologies in the Appalachians (USA). The challenge, however, continues to be obtaining a refined model that would allow producing an integral and accurate picture of weathering intensity and rate at the global level (e.g., Hartmann and Moosdorf 2011).

Glossary

Felsic: A silicate mineral or igneous rocks that are relatively rich in elements that form silicate minerals (e.g., feldspar). It also refers to those magma and rocks which are enriched in the lighter elements such as silicon, oxygen, aluminum, sodium, and potassium. They are usually light in color and have specific gravities less than 3. The most common felsic rock is granite.

Illitization: Transformation by leaching of a potassium-rich mineral into illite.

Lateritic (Adjetive of Laterite): Soil types rich in iron and aluminum, formed in hot and humid tropical areas. Nearly all laterites are rusty-red because of iron oxides. They develop by intensive and long-lasting weathering of the underlying parent rock.

Mafics: A silicate mineral or rock that is rich in magnesium and iron. Most mafic minerals are dark in color and the relative density is greater than 3. Common rock-forming mafic minerals include olivine, pyroxene, amphibole, and biotite. Common mafic rocks include basalt, dolerite, and gabbro.

Principal components analysis: Principal component analysis (PCA) is a mathematical procedure that uses an orthogonal transformation to convert a set of observations (matrix) of possibly correlated variables into a set of values of linearly uncorrelated variables called principal components. The number of principal components is less than or equal to the number of original variables.

Saprolite: A chemically weathered rock. It forms in the lower zone of soil profiles and represents deep weathering of the bedrock surface. In most outcrops its color comes from ferric compounds.

Sesquioxides: An oxide containing three atoms of oxygen with two atoms (or radicals) of another element. Many sesquioxides contain metals in the +3 oxidation state and the oxide ion, e.g., Al_2O_3, La_2O_3.

References

Amundson R (2005) Soil formation. In: Drever JI (ed) Surface and ground water, weathering and soils. Elsevier, Amsterdam

Bahlburg H, Dobrzinski N (2011) A review of the Chemical Index of Alteration (CIA) and its application to the study of Neoproterozoic glacial deposits and climate transitions. In: Arnaud E, Halverson E, Shields-Zhou (eds) The geological record of Neoproterozoic glaciations. Memoirs Geological Society, London

Boeglin JL, Probst JL (1998) Physical and chemical weathering rates and CO_2 consumption in a tropical lateritic environment: the upper Niger basin. Chem Geol 148:137–156

Braga MAS, Lopes Nunes JE, Paquet H et al (1990) Climatic zonality of coarse granitic saprolites ("arènes") in Atlantic Europe from Sandinavia to Portugal. Sci Géologique Memoire 85:99–108

Brantley SJ (2005) Reaction kinetics of primary rock-forming minerals under ambient conditions. In: Drever JI (ed) Surface and ground water, weathering, and soils. Elsevier, Amsterdam

Brantley SJ, Goldhaber MB, Vala Ragnarsdottir K (2007) Crossing disciplines and scales to understand the critical zone. Elements 3(5):307–314

Buggle B, Glaser B, Hambach U et al (2011) An evaluation of geochemical weathering indices in loess-paleosols studies. Quatern Int 240:12–21

Butman D, Raymond PA (2011) Significant efflux of carbon dioxide from streams and rivers in the United States. Nature Geoscience doi. doi:10.1038/NGEO1294

Chittleborough DJ (1991) Indices of weathering for soils and paleosols formed on silicate rocks. Aust J Earth Sci 38:115–120

Colman SM, Dethier DP (1986) An overview of rates of chemical weathering. In: Colman SM, Dethier DP (eds) Rates of chemical weathering of rocks and minerals. Academic Press, London

Cox R, Lowe DR, Cullers RL (1995) The influence of sediment recycling and basement composition on evolution of mudrock chemistry in the southwestern United States. Geochim Cosmochim Acta 59(14):2919–2940

Depetris PJ, Pasquini AI (2007) The geochemistry of the Paraná River: an overview. In: Iriondo MH, Paggi JC, Parma MJ (eds) The middle Paraná River: limnology of a subtropical wetland. Springer-Verlag, Berlin

Depetris PJ, Probst JL, Pasquini AI et al (2003) The geochemical characteristics of the Paraná River suspended load: an initial assessment. Hydrolog Process 17:1267–1277

Dosseto A, Buss HL, Suresh PO (2012) Rapid regolith formation over volcanic bedrock and implications for landscape evolution. Earth Planet Sci Lett 337–338:47–55

Ekosse GE, Ngole VM (2012) Mineralogy, geochemistry and provenance of geographic soils from Swaziland. Appl Clay Sci 57:25–31

Faure G (1998) Principles and applications of geochemistry. Prentice Hall, Upper Saddle River

Fedo CM, Nesbitt HW, Young GM (1995) Unraveling the effects of potassium metasomatism in sedimentary rocks and paleosols, with implications for paleoweathering conditions and provenance. Geology 23:921–924

Gaillardet J, Dupré B, Allègre CJ (1999) Geochemistry of large river suspended sediments:silicate weathering or recycling tracer? Geochim Cosmochim Acta 63:4037–4051

Goldberg K, Humayun M (2010) The applicability of the chemical index of alteration as a paleoclimatic indicator: an example from the Permina of the Paraná Basin, Brazil. Palaeogeogr Palaeoclimatol Palaeoecol 293:175–183

Goldich SS (1938) A study of rock weathering. J Geol 46:17–58

González-Alvarez I, Kerry R (2012) Weathering intensity in the Mesoproterozoic and modern large-river systems: a comparative study in the Belt-Purcell Supergroup, Canada and USA. Precambr Res 208–211:174–196

Hamadan J, Burnham CP (1996) The contribution of nutrients from parent material in three deeply weathered soils of peninsula Malaysia. Geoderma 74:219–233

Hartmann J, Moosdorf N (2011) Chemical weathering rates of silicate-dominated lithological classes and associated liberation rates of phosphorus on the Japanese Archipelago—Implications for global scale analysis. Chem Geol 287:125–157

Hofmann AW (1988) Chemical differentiation of the Earth: the relationship between mantle, continental crust and oceanic crust. Earth Planet Sci Lett 90:297–314

Krauskopf KB, Bird DK (1995) Introduction to geochemistry, 3rd edn. McGraw-Hill, New York

Harnois L (1988) The CIW index: a new chemical index of weathering. Sed Geol 55(3–4):319–322

Lecomte KL, Pasquini AI, Depetris PJ (2005) Mineral weathering in a semiarid mountain river: Its assessment through PHREEQC inverse modeling. Aquat Geochem 11:173–194

Li C, Yang S (2010) Is chemical index of alteration (CIA) a reliable proxy for chemical weathering in global drainage basins? Am J Sci 310:111–127

McLennan SM (1993) Weathering and global denudation. J Geol 101:295–303

McQueen KG, Scott KM (2009) Rock weathering and structure of the regolith. In: Scott KM, Pain CF (eds) Regolith science. Springer, Dordrecht

McQueen KG (2009) Regolith geochemistry. In: Scott KM, Pain CF (eds) Regolith science. Springer, Dordrecht

Nesbitt HW, Markovics G (1997) Weathering of granidioritic crust, long-term storage of elements in weathering profiles, and petrogenesis of siliciclastic sediments. Geochim Cosmochim Acta 61(8):1653–1670

Nesbitt HW, Young GM (1982) Early Proterozoic climates and plate motions inferred from major element chemistry of lutites. Nature 299:715–717

Ohta T, Arai H (2007) Statistical empirical index of chemical weathering in igneous rocks: A new tool for evaluating the degree of weathering. Chem Geol 240(3–4):280–297

Parkhurst DL, Appelo CAJ (2013) Description of input and examples for PHREEQC version 3—A computer program for speciation, batch-reaction, one-dimensional transport, and

inverse geochemical calculations. US geological survey techniques and methods. http://pubs.usgs.gov/tm/06/a43/. Accessed 13 May 2013
Parker A (1970) An index for weathering of silicate rocks. Geol Mag 107:501–505
Pasquini AI, Depetris PJ, Gaiero DM et al (2005) Material sources, chemical weathering, and physical denudation in the Chubut River basin (Patagonia, Argentina): implications for Andean rivers. J Geol 113:451–469
Pettijohn FJ (1941) Persistence of heavy minerals and geologic age. J Geol 49:610–625
Potter PE, Maynard JB, Depetris PJ (2005) Mud & mudstones. Introduction and overview, Springer
Pola A, Crosta G, Fusi N et al (2012) Influence of alteration on physical properties of volcanic rocks. Tectonophysics 566–567:67–86
Price JR, Velbel MA (2003) Chemical weathering indices applied to weathering profiles developed on heterogeneous felsic metamorphic parent rocks. Chem Geol 202:397–416
Roalset E (1972) Mineralogy and geochemistry of Quaternary clays in the Numedal Area, southern Norway. Nor Geol Tidsskr 52:335–369
Ruxton BP (1968) Measures of the degree of chemical weathering of rocks. J Geol 76:518–527
Shao J, Yang S, Li C (2012) Chemical indices (CIA and WIP) as proxies for integrated chemical weathering in China: inferences from analysis of fluvial sediments. Sedimentary Geol 265–266:110–120
Shen J, Algeo TJ, Hu Q et al (2013) Volcanism in South China during Late Permian and its relationship to marine ecosystem and environmental changes. Global Planet Change 105:121–134
Schettler G, Oberhänsli H, Stulina G, Djumanov JH (2013) Hydrochemical water evolution in the Aral Sea Basin. Part II: Confined groundwater of the Amu Darya Delta—Evolution from the headwaters to the delta and SiO2 geothermometry. J Hydrol 495:285–303
Strømsøe J, Paasche Ø (2011) Weathering patterns in high-latitude regolith. J Geophys Res. doi:10.1029/2010JF001954
Tardy Y (1971) Characterization of the principal weathering types by the geochemistry of waters from some European and African crystalline massifs. Chem Geol 7:253–271
Velbel MA (1985) Geochemical mass balances and weathering rates in forested watersheds of the southern Blue Ridge. Am J Sci 285:904–930
Velbel MA (1993) Constancy of silicate-mineral weathering-rate ratios between natural and experimental weathering. Chem Geol 105:89–99
Vogt T (1927) Sulitjelmafelets geologi og petrografi. Norges Geologiske Undersokelse 121:1–560
White AF (2005) Natural weathering rates of silicate minerals. In: Drever JI (ed) Surface and ground water, weathering and soils. Elsevier, Amsterdam

Chapter 6
The Wearing Away of Continents

Abstract Denudation is a substantial part of a cycle in which flowing water associates the terrestrial sector of the global hydrological cycle with continental wearing down, of which it is a chief agent. Tectonic uplift stimulates fluvial erosion, on contact with the water cycled by solar energy and clustered in surface channels by catchment processes. Progressive sediment transfers occur between upper and lower catchments, and subsequently between lower catchments and the marine environment. Big river flood plains store sediment in larger systems for longer periods, where reworking continues mechanical and chemical sorting before reaching the onward transfer of mature sediments to the coastal zone. This is a continuous process where coarse, raw fluvial sediments are eventually swept as *molasse* into trenches and back-arc basins, close to orogens. About 19.1 Gt of sediment and 3.8 Gt of dissolved phases are annually transferred to coastal oceans by river discharge. This huge amount of material is, directly or indirectly, the result of the action of weathering, which is a significant link in the cycling of carbon and, hence, a participant in controlling the Earth's climate. A significant portion of this material is, however, an indirect product of weathering because it is rock debris that has been recycled, having passed two or more times through the Earth's *exogenous cycle*. Also, anthropogenic activities are responsible for opposing actions, which increase the denudation rate through soil erosion, on one hand, and sequester sediments in human-made reservoirs, on the other.

Keywords Denudation · Erosion · Dissolved solids · Water chemistry · Suspended sediments · Material provenance · Sedimentary recycling · Climate change · CO_2 consumption · Runoff

6.1 Introduction

The synergistic action of rock weathering and erosion results in continental denudation; simply stated, continents are worn down by the disintegration of rocks, followed by the transport of the resulting debris and dissolved phases to

P. J. Depetris et al., *Weathering and the Riverine Denudation of Continents*, SpringerBriefs in Earth System Sciences, DOI: 10.1007/978-94-007-7717-0_6,

coastal seas and oceans. Rivers are the leading actors in the play because they constitute the major transport pathways; although they only represent about 0.0001 % of the total water mass existing on the surface of the Earth, they make up an effective conveyor belt system that transports over 19 Gt of sediment and 3.8 Gt (1 Gt = 10^9 t) of total dissolved solids (TDS) per year to the World's oceans, thus denuding the Earth's crust. These mechanisms of rock weathering and subsequent erosion are more complex than they appear at first sight because they not only wear away continents but also generate a rebound of the Earth's crust that interacts with climate constituting a mechanism that intervenes in the global carbon budget, affecting the concentration of atmospheric CO_2 (e.g., Ruddiman 1997).

6.2 The Exported Products of Weathering

The series of processes that *comminute* and dissolve crustal minerals and rocks, as seen in previous chapters, generate products that consist of dissolved and solid phases.

The naturally occurring dissolved fraction is composed of major (e.g., sodium, calcium, silicon, magnesium, alkalinity, sulfate, etc.), minor (e.g., iron, aluminum, manganese, etc.), and trace elements (e.g., the REE, transition metals). The concentrations of major elements are determined in the range of mg L^{-1} or mol L^{-1}; minor components are within the range of several parts per million (ppm) parts, and traces are determined in the <1 ppm (in the parts per billion parts or ppb range) or even less (e.g., ng L^{-1}). Some are soluble large cations, known as "large ion litophile" or LIL elements, such as strontium and cesium. Others are transition metals (e.g., scandium, titanium, vanadium, chromium) and high-field strength (HFS) elements, like niobium and thallium.

The solid fraction includes minerals (e.g., quartz, feldspar, micas, and heavy minerals); rock fragments; sometimes partially leached aluminosilicates (i.e., poorly ordered crystals); iron, manganese, and aluminum oxides and hydroxides; and crystalline *phyllosilicates* (i.e., the clay fraction).

It is not within the objectives of this brief monograph to enter the mineralogical complexities of the solid residue of weathering. The brief analysis that follows is limited to the most conspicuous characteristics of the material, whether solid or dissolved, that abandons the continents via riverine pathways.

6.2.1 Particulate Phases

The fractured material on the Earth's surface that results from different mechanisms (i.e., mechanical weathering is significant but tectonics is also responsible for the fracture of rocks), which is amenable to additional weathering processes, varies in size from boulders (i.e., up to ~4 m in diameter) down to cobbles

(256–128 mm), pebbles (64–8 mm), sand (2 to ~0.1 mm), silt (88 to ~4 μm), clay (~4–0.2 μm), and *colloids* (<~0.2 μm). Rock fragments and coarse rock debris are typical of high-energy systems (e.g., mountainous rivers) (Wohl 2010), whereas finer grain-size material (i.e., sand, silt, clay, and colloids) dominate in alluvial, mature rivers, with channels and floodplains that are self-formed in unconsolidated or weakly consolidated sediments. Rock fragments are generally composed of several minerals but sand and the finer grain-size fractions are usually made up by one dominant mineral phase.

Coarse geologic materials are mainly transported as *bed load* by high-energy rivers (also by mass movements like landslides and debris flows) and seldom reach the continental platform, although rivers draining active continental margins (e.g., along the Pacific coast of South America) are thought to be a significant conduit for the transfer of relatively coarse continental material to the sea (Milliman and Syvitski 1992). At any rate, relatively coarse grain-size sediments (i.e., sands) remain mostly along the continental–oceanic interface. Potter (1986) made an interesting study of beach sand composition in South America on the basis of the proportions of quartz, feldspar, and rock fragments (Q:F:Rf), concluding that tectonics are the dominant control on the mineralogy of beaches, showing a clear distinction between active and passive margins although, sometimes, active margin composition overprints a passive margin association, as it occurs along Argentina's Atlantic coast. The significant role of climate becomes evident in the so-called Brazil's Q:F:Rf association (91:4:5), which contrasts with the ratio found for the high-energy rivers draining South America's Pacific margin (24:16:60), showing the significance of climate in the former and the dominant role of mechanical weathering in the latter.

Finer grain-size sedimentary material (i.e., fine silt, clay, and colloids) is mostly transported as *total suspended solids* (TSS) by large rivers to estuaries, continental platforms and, eventually, to the deeper parts of world oceans after one or more cycles of deposition and transport (e.g., by ocean currents).For several decades, the mineralogy of deep-sea sediments has been associated with continental sources (e.g., Biscaye 1965; Griffin et al. 1968), mostly supplied by rivers and, to a smaller extent, as wind-transported dust and ice-rafted debris.

Over 15 years ago, Canfield (1997) produced an interesting study on the major elements geochemistry of river particulates from continental USA. Canfield (1997) combined a model that predicted river solutes as a function of precipitation, temperature, and a restricted number of weathering parameters. With another predictive equation, he found out that the composition of river particulates depends both on climate parameters (runoff and temperature, as they play an important role in defining dissolved river geochemistry) as well as on nonclimate factors, such as elevation, relief, tectonics, and drainage basin area. To the best of our knowledge, this approach has not been tested elsewhere.

The scientific research which has been carried out for many years, seeking to assess the global geochemical nature of mineral matter exported from the continents to the seas shows that, as river drainage basins grow larger, the mean composition of their suspended load approaches the composition of the upper

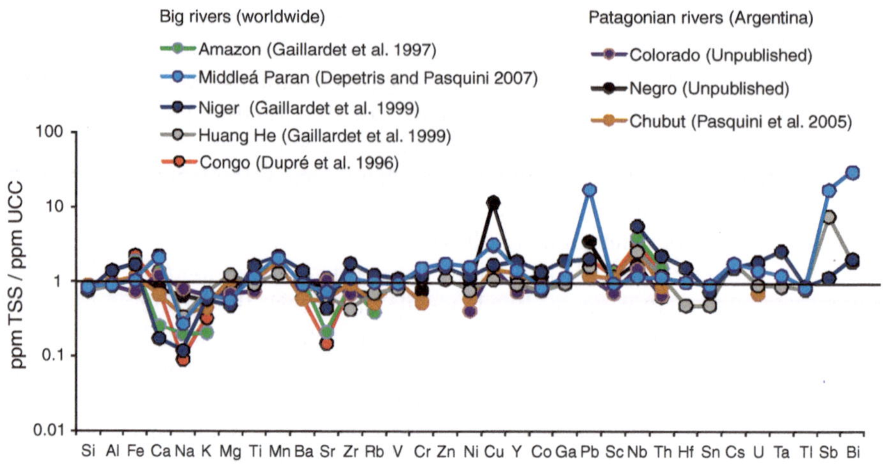

Fig. 6.1 Example of an extended UCC-normalized multi-elemental graph of TSS from major world and Patagonian rivers showing outstanding negative departures which are associated with elemental solubility (e.g., Ca, Na, K, Sr, and Rb), and positive deviations which are likely related with anthropogenic enrichment (e.g., Cu, Pb, and Sb) or insolubility (e.g., Fe, Ti, and Mn). Elements are ordered to obtain a monotonic decrease of UCC abundances when normalized to primitive mantle concentrations

continental crust (UCC). Extended UCC-normalized multielemental graphs of TSS from major world rivers (Fig. 6.1) show negative departures for soluble elements (e.g., Na, Rb, Sr, Ca, Mg, and K) and enrichments for insoluble (e.g., Ti, Fe, Zr) or adsorbed (e.g., Co, Cu, Pb, Cs, and REE) elements (e.g., Depetris et al. 2003). It is interesting to add that rivers appear to preserve a geochemical source signal in fine-grained TSS just a Potter (1986) found in South America's beach sands. The origin of sediments can be traced not only with isotopes (e.g., McLennan et al. 1993) but also with REE and other trace elements. The latter, for example, was the approach followed by Pasquini et al. (2005) when the geochemical study of the Chubut River was undertaken, finding that an active (i.e., volcanic arc) margin geochemical signature was persistent along Patagonia's passive Atlantic coast.

6.2.2 Dissolved Phases

The dissolved phases (also known as *total dissolved solids or TDS*) transported from the continents to the sea via rivers, groundwater, and wind-driven aerosols recognize three possible natural sources: weathering, volcanic manifestations, and ocean water recycled through atmospheric pathways. It is widely known that anthropogenic activities constitute an additional source, which may introduce human-made products in the environment or increase the flux of naturally occurring elements. Each source carries a chemical signature which, in the case of dissolved matter, is more difficult to establish than in solid phases, even in spite of the fact that the finer grain-sized material (e.g., clays and colloids) have suffered extreme modifications. The truly dissolved fraction begins with particles <1 nm (Gaillardet et al. 2005).

Meybeck (2005) has revisited the subject matter, updating many earlier studies. His main conclusions may be summarized as follows: (a) river water types are multiple; about a dozen water types can be identified according to the exposed lithology, water balance, and atmospheric input (Fig. 6.2); (b) Gibbs' (1970) global scheme (i.e., rain, weathering, and evaporation/crystallization dominance) still holds for about 80 % of river waters but does not account for the remaining 20 %, which has more complex driving variables; (c) few areas of the world still exhibit relatively pristine river geochemistry in as much as there is a global scale increase of riverine Na^+, K^+, Cl^-, and SO_4^{2-}. On a global basis, HCO_3^- concentration appears as stable (Meybeck 2005) although some large rivers, like the Mississippi, show for the last half century an amplification in the alkalinity export rate that has been linked to increased rainfall and to changes in the amount and type of land cover (Raymond and Cole 2003).

Contrasting with major constituents, trace elements in natural waters are characterized by concentrations <1 mg L^{-1}. There are several complex factors that control the concentration of dissolved trace elements, whether natural or anthropogenic, and dealing with such subject is beyond the scope of this monograph. We can only state here that their concentrations are several orders of magnitude lower than the host rocks that brought them to the Earth's surface (Fig. 6.3). Moreover, they exhibit an intricate dynamics which involves aqueous speciation, and adsorption onto colloids and other exchange surfaces, such as organic matter or hydrous oxides, which are the most outstanding features controlling their concentrations in the aquatic domain (Gaillardet et al. 2005).

6.3 Material Provenance: How Much is Recycled Material?

Provenance is a term that comes from the French "provenir," and means "to come from." In the Earth Sciences, provenance concerns the lithologic source of a rock, usually in sedimentary rocks. It does not refer, as one would think, to the circumstances of the collection of a rock sample. In dealing with continental denudation by the action of rivers, the term is used to establish the source of the sediment load transported to coastal seas.

Most mountainous regions (e.g., the Andes, the Himalayas) have a high fraction of outcropping sedimentary rocks, which raises the potential for sedimentary-sourced particles to dominate the TSS of major rivers. Fine-grained sedimentary rocks are frequently the product of the turnover of preexistent clastic sediments because their particles mostly consist of stable weathering products, and these

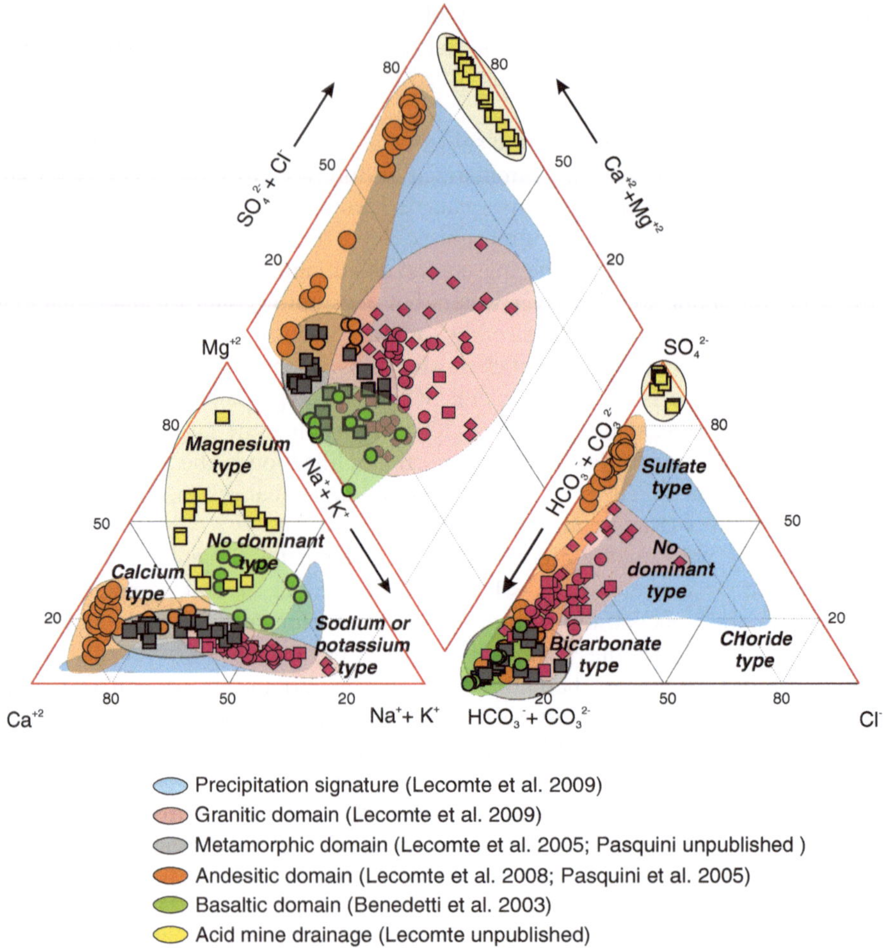

Fig. 6.2 Piper diagram showing the chemical clusters determined by different dominant lithological compositions. The Piper classification is based on the major chemical composition of natural waters

particles can be recycled through several episodes of burial, uplift, and erosion. The recycled condition of the TSS in major rivers has been recognized for many years but it is difficult to quantify. Veizer and Jansen (1979) estimated that, on a global basis, clastic sedimentary rocks are 65 % recycled. Notwithstanding, sediment recycling is a complex subject and the reconstruction of the burial and exhumation cycles, particularly in areas with elaborate tectonics is a problem that requires detailed analyses. As an example, Garzanti et al. (2013) have explored the complexities of sediment recycling at convergent plate margins (Indo-Burman ranges and Andaman-Nicobar ridge).

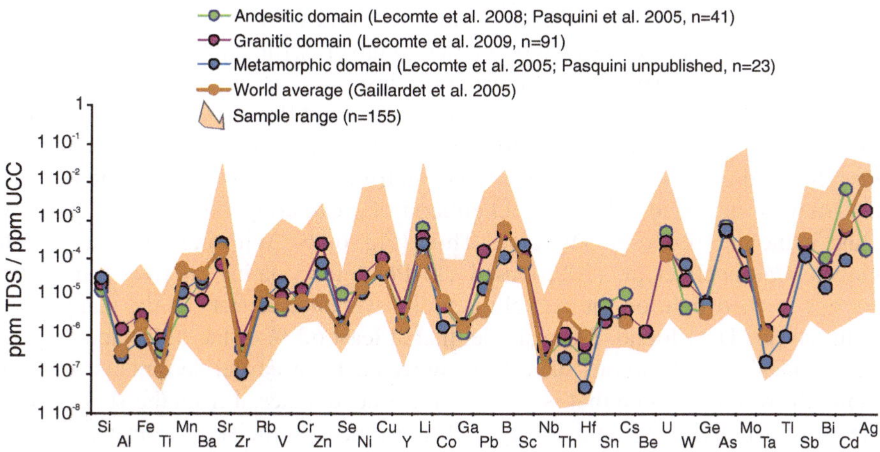

Fig. 6.3 Example of an extended UCC-normalized multielemental graph of TDS for different dominant lithological compositions. Shaded area is determined by maximum and minimum normalized concentration for each element. UCC-normalized concentrations mostly range from 10^{-2} to 10^{-7} and are mainly controlled by mineral solubility and adsorption processes

In most major rivers, there is a surplus of TSS over what would be expected from the dissolved load if only crystalline rocks are being weathered and eroded. The amount of this excess is proportional to the extent of sedimentary rock cover in the source area. This approach provides an estimate of the fraction of recycled material a river is carrying. Gaillardet et al. (1997) used this line of thought to complete a study on the chemical and physical denudation in the Amazon River basin. Later, Gaillardet et al. (1999) attempted to estimate the proportion of recycling in different drainage basins by comparing the measured TSS to that expected from the composition of the dissolved load. The results vary broadly for different rivers with values mostly falling in the range of 45–75 % recycled material. Using the described methodology, Gaillardet et al. (1999), for instance, anticipated significant recycling for the Paraná River drainage basin TSS load, estimating that about 14 % of the eroded material is supplied by pristine continental sources.

Chemical indices, as seen in Chap. 5, can assist in establishing the likelihood that a river-transported sediment load has a significant proportion of recycled material. If, for example, CIA is high and the ICV is low, then the cause could be either extensive weathering in the source area or significant recycling of previously weathered material. Conversely, if a particular TSS exhibits low CIA and high ICV, it is very likely that the parental source is composed of mostly poorly weathered crystalline rocks.

For over 20 years, geochemical and isotopic approaches to constrain sediment provenance have supplied tools that proved to be the most valuable to attain such goal. An outstanding example is the work of McLennan et al. (1993) that used trace elements (notably REE) and a myriad of isotopes (mainly from the U/Pb system) to collect fingerprints for several terrain types, recycled sedimentary rocks among them. $^{87}Sr/^{86}Sr$ and $^{143}Nd/^{144}Nd$, often used with the *epsilon (ε) notation*, have also proved powerful methodologies to unveil sediment sources, not only in river systems, but also in wind-transported dust.

For example, we can cite the work of Depetris and Pasquini (2007), who used a Zr/Sc versus Th/Sc bivariate plot to show the possibility of significant recycling in the TSS load of upper Paraná River Andean tributaries (Fig. 6.4a). Another graph that may be useful to exhibit the likelihood of weathering and sedimentary recycling is the one that used Th versus Th/U (Fig. 6.4b). With more powerful techniques, Campbell et al. (2005) used He–Pb double dating in river-transported zircons to shed light on the recycling and provenance aspects of the Ganges and Indus rivers. They found, for example, that at least 60 % of the Indus and 70 % of the Ganges detrital zircons have been recycled from earlier sediments. Furthermore, the results implied that ~75 % of the eroded material from the Himalayas is derived from areas of anomalously high erosion, with the short-term exhumation rate exceeding the long-term average.

6.4 Weathering, CO_2 Consumption, and Climate Change

As seen in Chap. 4, the action of weathering is an effective mechanism to sequester CO_2. In a worthwhile revisit of this aspect we can recur to the following example; in the presence of carbonic acid, the Mg-olivine forsterite decomposes into:

$$Mg_2SiO_{4(s)} + 4CO_{2(g)} + 4H_2O \rightarrow 2Mg^{2+}_{(aq)} + 4HCO^-_{3(aq)} + H_4SiO_{4(aq)} \quad (6.1)$$

Four moles of CO_2 are taken from the bacteria-supplied soil air, of which two are immobilized when Mg^{2+} is precipitated as magnesite:

$$2Mg^{2+}_{(aq)} + 4HCO^-_{3(ac)} \rightarrow 2MgCO_{3(s)} + 2CO_{2(g)} + 2H_2O \quad (6.2)$$

It is clear then that for every mole of weathered forsterite, two moles of CO_2 from the gaseous pool become fixed in the form of carbonate and the other two CO_2 mol return eventually to the atmosphere. The implications of this simple geochemical mechanism have been treated extensively by many authors (e.g., Edmond and Huh 1997), underlining the role played by the weathering-connected carbon dioxide in the control of the Earth's climate.

The described mechanism is only a link in the chain of climate control. For example, increased erosion of high-elevation regions causes a rebound (*isostasy*) of the underlying rock layers in response to overburden removal. However, the resulting uplift compensates only partially the height-loss due to erosion, thus leaving the original Earth's surface lower than it had been. This system driven by world tectonics has been also linked to climate change (e.g., Ruddiman 1997) and supplies an intricate vision of the scenario that surrounds continental denudation.

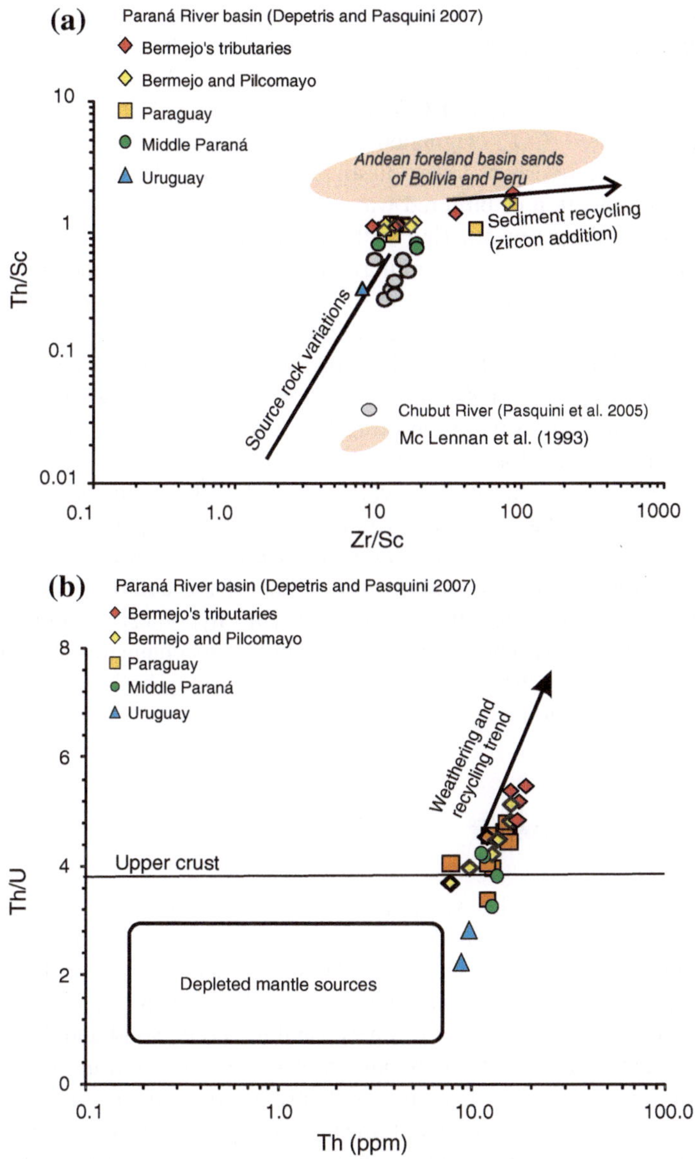

Fig. 6.4 **a** Th/Sc ratio and Zr/Sc ratio increase from mafic to felsic source areas. The anomalous Zr is related to recycling of older sediments. Data from the indicated sources; some Paraná tributaries (Bermejo and Paraguay) show the likelihood of recycling. **b** Plot of Th/U versus Th for TSS of Paraná River and tributaries. During weathering, there is an increase trend of Th/U above UCC values (3.5–4.0). A similar example was shown by McLennan et al. (1993)

It is clear then that weathering consumes CO_2 (e.g., Boeglin and Probst 1998) and, as long as there is an association with a viable mechanism, it may insure the sequestration of carbon.

If, in contrast, the evasion of CO_2 from inland waters is not significantly hindered, then respired CO_2 originated largely from terrestrial ecosystems (i.e., introduced into the hydrosphere as dissolved soil CO_2), the oxidation of organic carbon, the acidification of buffered waters, etc., determines a significant efflux of CO_2. Modeling mineral weathering with the PHREEQC computer code in a high mountainous river (Córdoba, Argentina) has shown the significance of CO_2 in the overall weathering mechanism: in the stream's uppermost reaches, CO_2 accounts for nearly 57 % of the total phases intervening in the process (Lecomte et al. 2005). Adding significantly to the whole picture, recent research has shown that a large set of streams and rivers in the USA are supersaturated with CO_2 when compared with the atmosphere, emitting 97 ± 32 Tg (1 Tg $= 10^{12}$ g) of carbon each year (Butman and Raymond 2011).

The sequestration of atmospheric carbon has played a main role in the theories that propose such mechanism as the trigger that determined periods with long-term CO_2 changes (i.e., with low temperatures). Such is the case of the hypothesis identified with an acronym based on the initials of the authors (BLAG) (Berner et al. 1983) and, years later, the so-called Raymo hypothesis (Raymo and Ruddiman 1992). Both hypotheses propose dissimilar mechanisms to control low-temperature periods in the geological past although they both assign weathering and carbon sequestration a relevant role.

6.5 Continental Runoff

Runoff is understood as the water flow that occurs when the surface (i.e., soil, sediment, or rock) is permeated to its full capacity and the excess water from rainfall, melt water, or other sources flows over the land. Runoff is relatively short-lived and soon ends reaching the channel, which is the passageway of a river. Responsive to variations in water discharge and load (i.e., solid and dissolved), the channel constantly adjusts its shape and course, thus constituting a dynamic element of the landscape. Runoff is thus transformed into stream flow.

A river's channel is an efficient means for running water. Hence, the first requirement that affects runoff and stream flow within a basin would be the supply of water, making climate an important item in any river scenario. It eventually determines not only the water supply through atmospheric precipitation, but also the degree to which that precipitation is returned to the atmosphere as water vapor before it can contribute to stream flow. Climate is considered to include not only the long-term moisture supply, but also the daily weather patterns that establish the timing and total quantity of water supplied to a drainage basin. Controlling weather variability there are intervening factors such as storm types, typical duration and intensity of rainfall, frequency, distribution of precipitation over the basin, kind of precipitation (rainfall or snowfall), and reliability of precipitation over a period of years. Periodic floods are frequently influenced by physical conditions within the drainage basin, but droughts are controlled entirely by climate.

The riverine network that dissects the Earth's surface provides the key linkage between land and sea, delivering yearly about 36,000 km^3 of freshwater to the coastal seas. Although impressive, this enormous volume represents only ~0.0001 % of the total water at the Earth's surface. As much as 40 % of total riverine discharge is supplied only by a dozen big rivers, like the Amazon, Congo, Orinoco, Changjiang, etc.

Climate, volcanism, modifications in the relative base level but, above all, plate tectonics are jointly responsible for the location, size, shape, and orientation of big river systems. Most of today's large river systems in Eurasia and the Americas date from the Miocene; some are much older and originated in the Mesozoic and some as far back as the Carboniferous. A few, such as China's Yellow and Yangtze are thought to have a Pliocene–Pleistocene age, and the present Nile was formed in the Pleistocene. The Paraná and Uruguay rivers, for example, are thought to have originated during the Late Cretaceous but the Paraguay River, a significant tributary to the former, is a younger feature, with likely Mid-Tertiary age (Potter and Hamblin 2006).

Having examined briefly the question of the geological age of the current riverine network, we can now return to the aspect of continental runoff. River discharge is—as stated above—a direct consequence of climate variability and the size of the drainage basin. Data collected from 1,100 world rivers have shown that, as expected, large drainage basins usually have higher discharges (Milliman and Farnsworth 2011). Similar drainage areas and comparable climate often exhibit similar discharge. For example, in Argentina's arid Patagonia, the Deseado (14,450 km^2) and Coyle (14,600 km^2) rivers have equivalent mean discharges (~5 $m^3\ s^{-1}$) (Depetris et al. 2005). There are, however, other examples that show the significance of the reverse: in comparable drainage basin areas but contrasting climate, discharge can vary by two to three orders of magnitude (Milliman and Farnsworth 2011).

Let us use South America's Paraná River as an example to illustrate some peculiarities of big rivers' hydrology. The Paraná River, in terms of drainage area, is the fifth largest basin in the world (~2.6×10^6 km^2). Near its mouth, the Paraná joins the Uruguay River to make up the Río de la Plata drainage basin. Added together the two drainage systems reach a total area of 3.17×10^6 km^2 and a mean discharge of ~21,500 $m^3\ s^{-1}$ (Pasquini and Depetris 2007 and references therein).

With ~180 mm y^{-1}, Paraná's runoff is relatively low for global standards (the Amazon's runoff is ~1,000 mm y^{-1}), and its current mean discharge is ~500 $km^3\ y^{-1}$ (~17,000 $m^3\ s^{-1}$), a water flow that places the Paraná within the World's twelve topmost discharges. Currently, most of Paraná total annual discharge (~63 %) is supplied by the upper Paraná River, which has its uppermost catchments near the Brazilian city of Sao Paulo. Figure 6.5a shows the over 100 years-long (1904–2011) mean monthly discharge time series at the gauging station of Corrientes, ~1,200 km upstream the river mouth. There are three features which become promptly apparent: (a) the series shows a marked seasonality

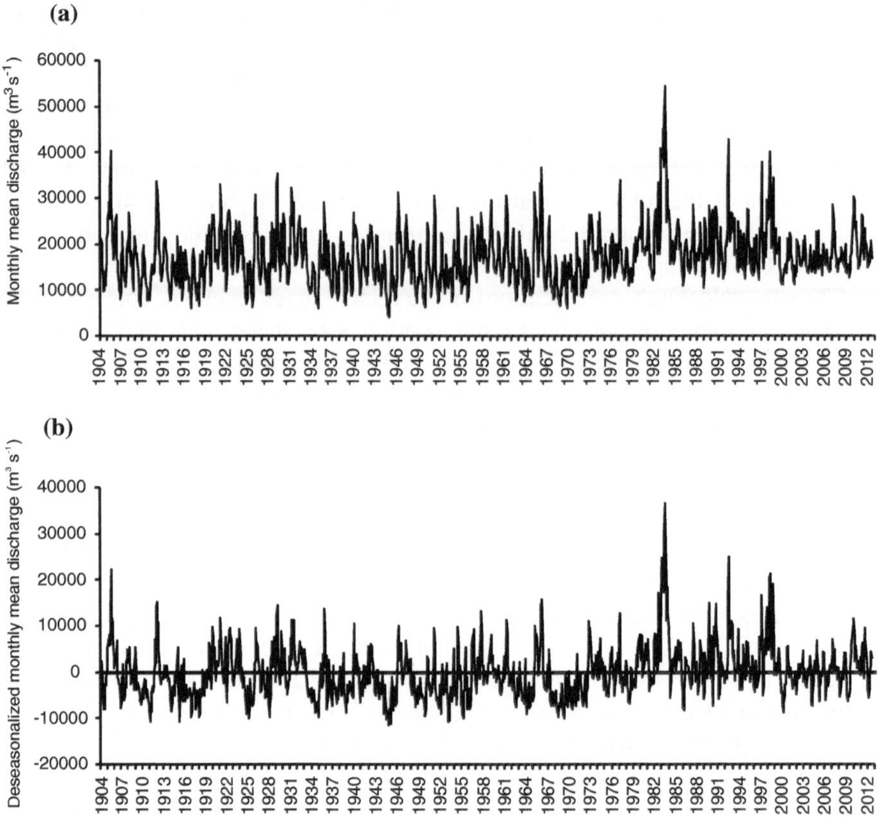

Fig. 6.5 **a** Paraná River monthly mean discharge series at Corrientes (~1,200 km upstream from the mouth) for the period 1904–2012. The series is characterized by seasonality and unusual discharge events. **b** The same as **a** but deseasonalized. Positive departures are frequently associated with El Ñiño events whereas negative are often linked with La Niña episodes

(i.e., high waters occur during the austral summer and low waters during winter); (b) there is a pronounced variability in mean monthly discharges; (c) interspersed in the series there are outstanding flood events. Since the series exhibits a regular seasonal fluctuation then, for the purpose of analysis (i.e., to estimate an underlying trend) it is necessary to remove the seasonality to produce a *deseasonalized* dataset. In the Paraná, the series with suppressed seasonality shows marked positive departures particularly noticeable in the period after 1970 and what appears to be more frequent negative departures prior to that year (Fig. 6.5b).

The 1904–2012 deseasonalized series can be used to scrutinize for recurrent signals by means of a *harmonic analysis* technique. In this case, as in earlier instances (e.g., Pasquini and Depetris 2007; Depetris and Pasquini 2008), we used the *continuous wavelet transform* (CWT) approach, which has the advantage of localizing the time-scale of a signal, revealing trends, breakdown points, and

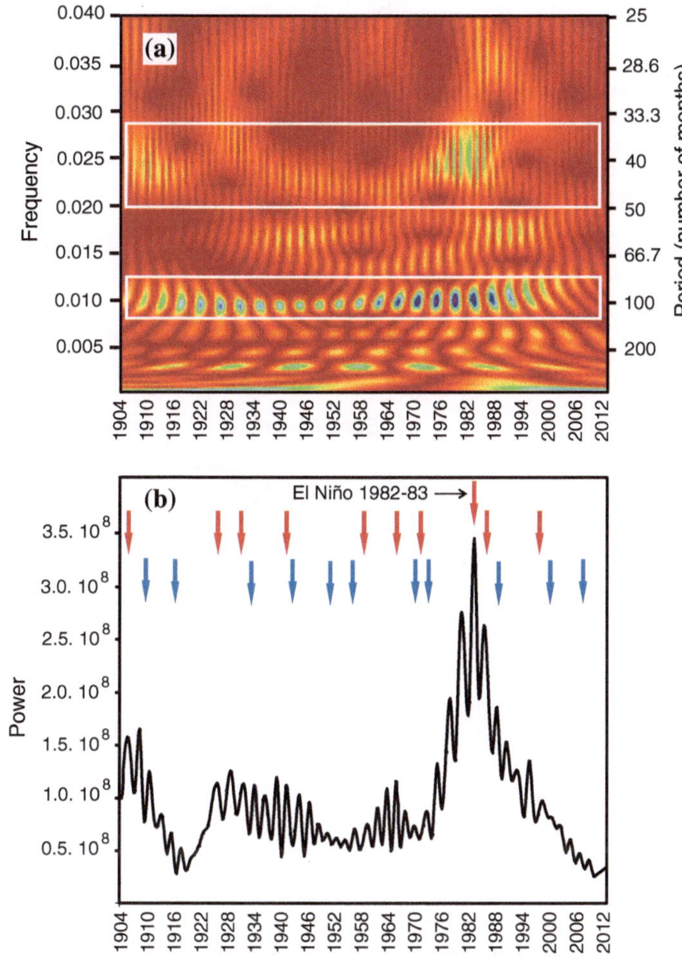

Fig. 6.6 a Real part of the continuous Morlet wavelet spectrum of the Paraná River deseasonalized monthly mean discharge at Corrientes (1904–2012); the framed areas show the occurrence of decadal and interannual periodicities in ENSO frequencies. **b** Wavelet power frequency-range for the 2–6 years frequency band that shows outstanding increased integrated power for the 1982–1983 and 1997–1998 El Niño events. Very strong El Niño episodes are identified with *red arrows* and strong La Niña events with *blue arrows*

discontinuities (e.g., Nakkem 1999). Figure 6.6a shows the wavelet spectra for the 2 to 6-year period, computed for the deseasonalized data of Fig. 6.5b. The power spectra show a clear signal supplied by high transform coefficients (power) for quasi-decadal periods and, also, in the *El Niño-Southern Oscillation* (ENSO).

Figure 6.6b shows the power spectra for the 2 to 6-year frequency range, computed for the deseasonalized Paraná River discharge series (Fig. 6.5b). Clearly, this range includes the period of recurring ENSO events, exhibiting the impact of the strong 1983 and 1997 occurrences.

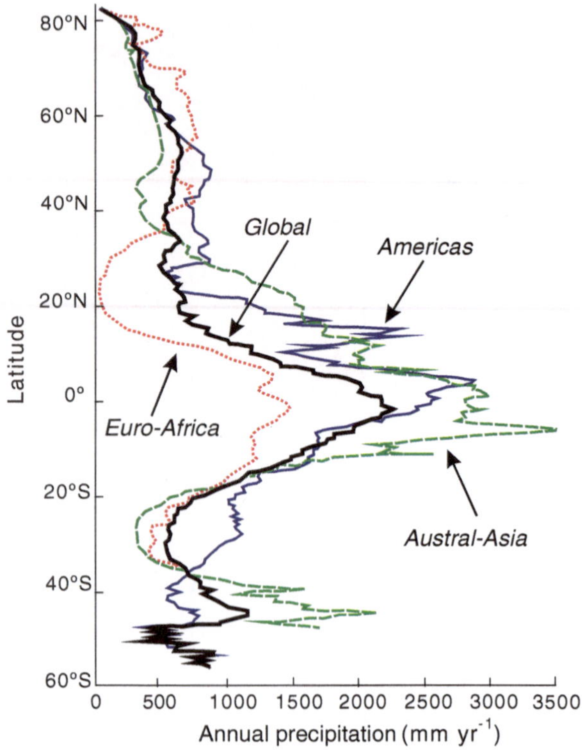

Fig. 6.7 Latitudinal allocation of global precipitation (*bold line*) and in different continents. Note the substantially lower Euro-African precipitation in the ~20° N–20° S latitudinal band (Milliman and Farnsworth 2011). Reproduced with permission, Cambridge University Press

It is not within the objectives of this monograph to enter the detailed analysis of global runoff (or river discharge) going beyond the illustrating examples used above. Milliman and Farnsworth (2011) have produced an updated, expanded, and valuable state-of-the-art. As a sum up, the global distribution of precipitation determines that South America's large rivers and those draining Southeast Asia and Oceania account for almost 70 % of the freshwater discharge to the continental coasts (Fig. 6.7; Milliman and Farnsworth 2011).

Rivers must be examined within the context of their history. This is the reason why relatively long series of discharge data are useful to assess discernible changes in their mean trend; the longer the series, the more confidence in the statistically significant results. Let us again use information recorded in the Paraná River to examine if its deseasonalized series shows a significant trend. Figure 6.8 shows the trend analysis performed by means of the Mann–Kendall methodology (also known as Kendall τ), which clearly shows a statistically significant positive slope, thus indicating that mean annual discharges have been increasing consistently, particularly during the last 50 years (Pasquini and Depetris 2007).

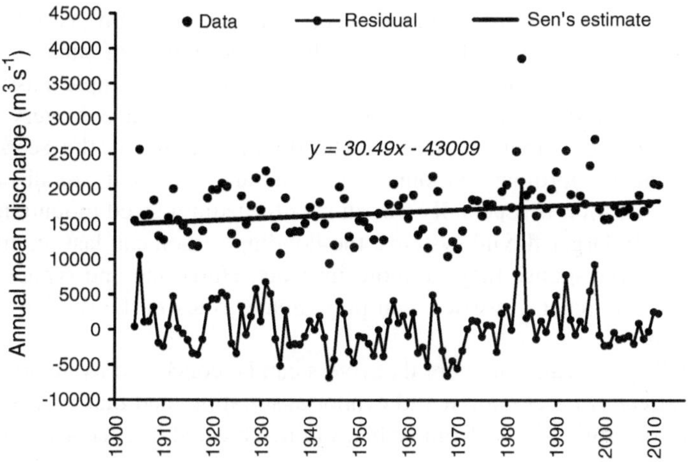

Fig. 6.8 Mann–Kendall trend test for data in Fig. 6.5 (1904–2012). Paraná's mean annual discharge has increased ~18 % during the last 100 years at a mean rate of ~30 $m^3 y^{-1}$

The seasonal Kendall test was employed to examine monthly trends. Table 6.1 shows the levels of significance that identify increasing discharge trends for the July–December period (i.e., austral winter and spring). It can be stated, then, that the Paraná's overall increasing discharge trend is basically accounted for by the increasing precipitation in the drainage basin that occurs in the austral winter and spring. The rest of the year, the river discharge does not show an apparent modification of its regime.

Table 6.1 Seasonal Kendall test results of the Paraná River monthly mean discharge time series (1904–2012) at Corrientes

Month	No. of data	Kendall τ	*p* value	Conf. level
January	109	0.4818	0.31497	NS
February	109	0.7715	0.21914	NS
March	109	−0.7715	0.21914	NS
April	109	0.1519	0.43964	NS
May	109	1.6104	0.05365	NS
June	109	1.6025	0.05451	NS
July	109	2.6631	**0.00387**	**99 %**
August	109	4.3468	**6.69×10^{-6}**	**99.9 %**
September	108	4.0141	**2.98×10^{-5}**	**99.9 %**
October	108	3.4831	**0.00024**	**99.9 %**
November	108	3.2336	**0.00061**	**99.9 %**
December	108	2.3681	**0.00894**	**99 %**

Note the statistically significant increasing trend (bold coefficients) between July and December
NS not significant

After considering unusual climatic events that may determine very high or very low runoff, a query immediately arises: what is the impact of such extraordinary events on the wearing down of continents? There is no simple answer to this question. In some fluvial systems, episodic events can occur over very short time intervals, frequently hours to weeks. The direct example is the seasonal flood during which extraordinary amounts of water and sediment are discharged by rivers. This situation is especially true in small- medium-sized mountainous rivers and streams. In larger fluvial systems, the flooding season can last months and the linkage process-response may be more intricate. Hurricane and typhoons, due to their catastrophic characteristics, can play greater roles in the discharge of water and sediment.

Concluding, we can state here that rivers can be conceived as integrators of the opposing effect of precipitation and evapotranspiration, and that they are dynamic systems whose runoff is far from stable, often subjected to events that profoundly impact on their regimes.

6.6 Continental Denudation

Denudation comes from "denudare," the Latin word that means "to strip of all covering, to lay bare." In the Earth Sciences, it is the term used to identify the long-term effect of all the exogenous processes added together, which cause the wearing away of the Earth's surface, leading to a decrease in elevation and relief of landforms and landscapes. Endogenous processes such as plate tectonics, and the related volcanic activity and earthquakes, uplift and exhume continental crust exposing it to processes linked to denudation, such as weathering, erosion, and mass wasting. As seen in previous chapters, the action of mineral and rock weathering results in a solid and dissolved residue and we will consider now the continent-ocean transfer separately.

The most updated database available to compute **solid continental denudation** via riverine transport to the Earth's oceans, supplies a mass transport rate of $\sim$19.1 Gt y^{-1} (or 19.1 10^{9} t y^{-1}), which can be translated to a global mean sediment yield of 190 t km^{-2} y^{-1} (or $\sim$0.1 mm y^{-1}) (Milliman and Farnsworth 2011). Earlier computations arrived at 18.3 Gt y^{-1} (Holeman 1968), 13.5 Gt y^{-1} (Milliman and Meade 1983), and 16 Gt y^{-1}, using an empirical approach (Ludwig and Probst 1998). These calculations are based on numerous determinations of the mass of sediments transported in suspension (TSS) or as bed load along the river bottom. The TSS load includes the wash load, which is the fine grain-size fraction that is mostly transported in suspension, and the bed material load, which is the sediment incorporated into the TSS during higher discharge. As river flow decreases, the sediment that is normally transported as bed material load may be incorporated to the river bed load. The opposite is also true: increased flow may incorporate a fraction of the bed load into bed material load.

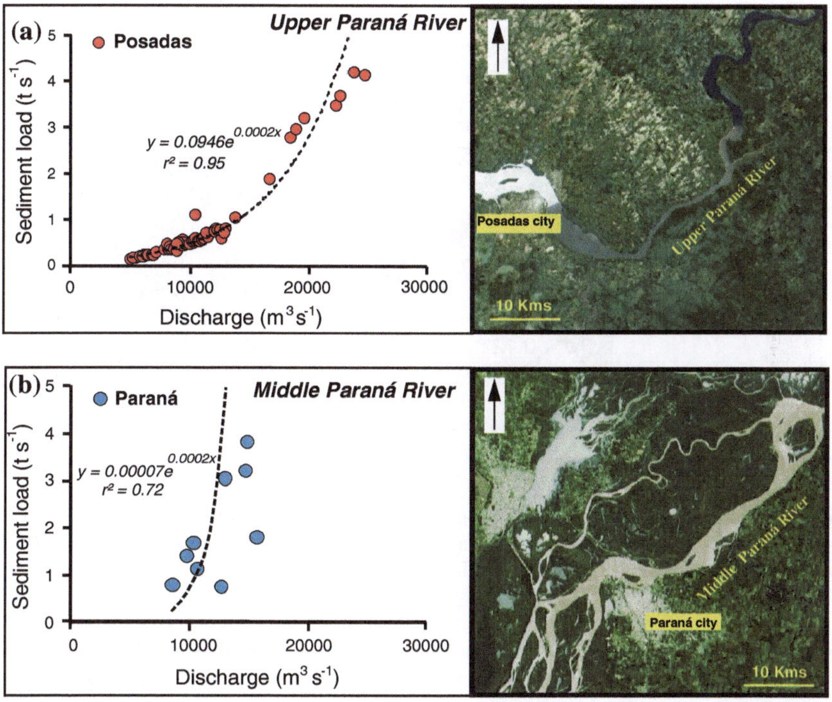

Fig. 6.9 **a** Relationship between discharge and sediment transport rate in the Upper Paraná River at Posadas (~1,500 km upstream the mouth); note in the Google Earth image the narrow flood plain, with high banks, and confined river channel. Measured suspended sediment concentrations can be correlated with river discharge to derive a sediment-rating curve. **b** Same as **a** but at Paraná (~400 km upstream the mouth); note the channel braided pattern and the wide (~30 km) flood plain, with ponds and secondary channels

Measuring the transport rate of the sediment bed load in a river, particularly if it is large, is a difficult task with traditional invasive methodology (i.e., sampling devices that are placed on the river bottom). Modern techniques, such as acoustic Doppler current profiler (ADCP), have arrived to supply a much less intrusive approach. However, most workers still assume that the bed load fraction of the total sediment transported by rivers is relatively small (e.g., ~10 %), although this relative amount may be too high for large meandering rivers or too low for steep, mountainous rivers, where the bed load mode of sediment transport operates almost exclusively during extreme flooding events.

Some of the complexities that surround obtaining an accurate measure of the sediment delivered by large rivers to the coastal ocean are illustrated in Fig. 6.9. A well-defined, confined channel, as in the Upper Paraná River (Fig. 6.9a) allows determining a much more reliable relationship, in statistical terms, between discharge and sediment transport rate. In a river section as the one shown in Fig. 6.9b, the mentioned relationship tends to be blurred by the water exchange that occurs

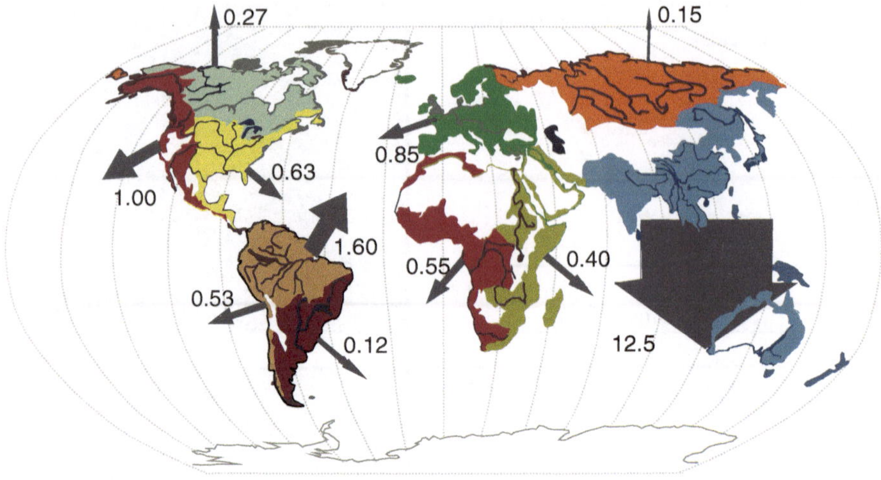

Fig. 6.10 Annual discharge of TSS to the global coastal Ocean. Fluxes in Gt y^{-1}. The total sediment mass is ~19 Gt y^{-1}. The figure was obtained from Milliman and Farnsworth (2011). Reproduced with permission, Cambridge University Press

between numerous side channels, the wide flood plain with ox-bow ponds and other morphological features, on one hand, and the river's main stem, on the other.

Again following Milliman and Farnsworth (2011), the above-stated global sediment flux of 19.1 Gt y^{-1} is distributed, in a descending order, as follows: Oceania ~37 %; Asia ~28 %; South America ~12 %; North/Central America ~10 %; Africa ~8 %; Europe ~4 %; and Eurasian Arctic ~1 %. These results underline the role of precipitation (i.e., climate) and relief (i.e., partly a surrogate for tectonics) in defining significant denudation and showing, as well, the linkage with weathering (Fig. 6.10).

As seen in earlier chapters, weathering generates a solid and a dissolved residue. Rivers, therefore, transport not only particulate matter but also the TDS that result from the action of weathering. In a schematic representation of the weathering reaction of a plagioclase, like the one seen in Chap. 4 (e.g., Na-plagioclase + water + carbonic acid → kaolinite + sodium ion + alkalinity + dissolved silica), a significant percentage (e.g., 20–30 %) of the initial reactants are transferred to the solution and, therefore, exported from the system. Therefore, the appraisal of denudation implies the consideration of the dissolved solids that are removed from the continental mass and transported to world oceans.

There has been considerable debate over the supremacy of climate over lithology (or the other way around) in the control of the riverine chemical signal. However, the idea that lithology is the factor that rules the nature and quantity of the TDS efflux is currently gaining ground. Climate, on the other hand, appears to control the rate of chemical weathering and, hence, of dissolved denudation.

World wide data gathered on the total dissolved mass transport rate shows that rivers discharge annually about 3.8 Gt of dissolved phases to the ocean, the

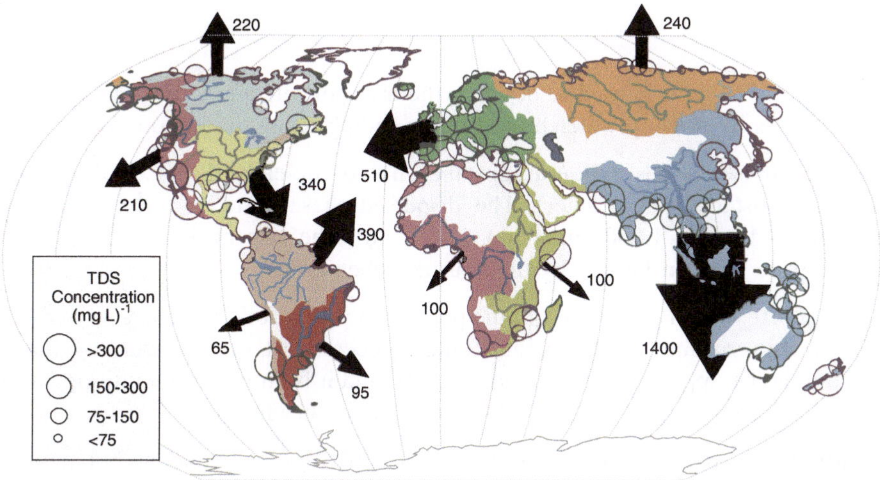

Fig. 6.11 Riverine delivery of TDS to the global coastal ocean. Numbers are mean concentrations in mg L^{-1}. *Circles* represent mean concentrations for selected rivers. Estimated global continental flux is ~3.8 Gt y^{-1}. More details are provided in the book by Milliman and Farnsworth (2011). Reproduced with permission. Cambridge University Press

greatest contribution being delivered by Asian rivers, which supply ~37 % of the global flux (Fig. 6.11). Europe occupies the second place with ~13 %, and the Amazon runs third with ~10 % (Milliman and Farnsworth 2011).

Before concluding the abridged treatment of continental denudation, we must consider, albeit briefly, the anthropogenic role on the denudation scenario. It is known that human activities have concurrently increased the sediment transport by global rivers through soil erosion—estimated by Syvitski et al. (2005) in 2.3 ± 0.6 Gt y^{-1}- and, at the same time, decreased the sediment transfer of particulate material to ocean coastal areas by retaining 1.4 ± 0.3 Gt y^{-1} in reservoirs (Syvitski et al. 2005). The authors have estimated in this significant paper, that over 100 Gt of sediment and 1–3 Gt of carbon are presently sequestered in reservoirs worldwide, which were mostly built within the past 50 years.

Sediment loads and yields have been frequently translated into denudation rates (e.g., Judson and Ritter 1964). Presupposing an average specific gravity of 2.0 for the upper portion of the crust that is subjected to erosion, a sediment yield of 2,000 t km^{-2} y^{-1} is approximately equal to a denudation rate of 1 mm y^{-1}. Therefore, yields that fluctuate between 100 and 10,000 t km^2 y^{-1} can be translated into 0.05 and 5 mm y^{-1}. These rates, expressed as a uniform decrease in height of the landscape, do not mean, naturally, that denudation occurs in this manner, but it is a simple way to compare rates.

Glossary

Bed load: The term is used to describe the transport of sand, gravel, boulders, or other debris by flowing water by rolling or sliding along the bottom of a stream.

Colloid: It is a solid or liquid substance microscopically dispersed throughout another substance (e.g., water). The dispersed-phase particles have a diameter between ~1 and ~1,000 nm (1 nm = 10^{-9} m). The dispersed-phase particles or droplets are affected largely by their surface chemistry.

Comminute: From *comminution*; the process in which solid materials are reduced in size, by crushing, grinding, and other processes. It occurs naturally during faulting in the upper part of the Earth's crust, and it is an important unit operation in mineral processing, ceramics, electronics, and other fields.

Continuous wavelet transform (CWT): It provides a redundant but detailed description of a signal in terms of both time and frequency. It is used to divide a continuous-time function into wavelets. Contrasting with Fourier transform, the CWT is characterized by the ability to construct a time–frequency representation of a signal that offers very good time and frequency confinement.

Deseasonalize: In statistics, data are deseasonalized when the regular seasonal fluctuations are removed from a time series.

El Niño-Southern Oscillation (ENSO): It is a band of anomalously warm ocean water temperatures that occasionally develops off the western coast of South America and can cause climatic changes across the Pacific Ocean. The Southern Oscillation refers to variations in air surface pressure in the tropical western Pacific and in the temperature of the surface of the tropical eastern Pacific Ocean (warming and cooling known as *El Niño* and *La Niña*, respectively). The two variations are attached: the warm oceanic phase, El Niño, occurs with high air surface pressure in the western Pacific, while the cold phase, La Niña, accompanies low air surface pressure in the western Pacific.

Epsilon (ε) notation: It is an alternative way of expressing isotope ratios which allows greater flexibility in the way in which isotopic data are presented; the value is a measure of the deviation of a sample or sample suite from the expected value in a uniform reservoir and may be used as a normalizing parameter for samples of different age. It is normally calculated for $^{143}Nd/^{144}Nd$ ratio and is used to represent parts in 10,000 by the following equation:

$$\varepsilon^{143}\mathrm{Nd} = \left[\frac{\left(\frac{^{143}\mathrm{Nd}}{^{144}\mathrm{Nd}}\right)\mathrm{sample}}{\left(\frac{^{143}\mathrm{Nd}}{^{144}\mathrm{Nd}}\right)\mathrm{standard}} - 1\right] \times 10{,}000$$

Exogenous cycle: It is a set of events or processes which are completed, returning to its beginning and then repeating itself in the same sequence. It describes the fluxes of materials, water, and gasses that occur at the intersection of the lithosphere, biosphere, hydrosphere, and atmosphere.

Harmonic analysis: A branch of mathematics concerned with the representation of functions or signals as the superposition of basic waves, and the study of and the generalization of the notions of Fourier series and Fourier transforms, harmonic functions, trigonometric series, almost periodic functions, and others.

Isostasy: The term is used in geology to refer to the state of gravitational equilibrium of all large portions of Earth's lithosphere as though they were floating on the denser underlying layer, the asthenosphere, a section of the upper mantle composed of plastic rock that is about 110 km below the surface.

Molasse: The term refers to the sandstones, shales, and conglomerates formed as terrestrial or shallow marine deposits in front of rising mountain chains. These deposits are typically the nonmarine alluvial anyd fluvial sediments of lowlands, as compared to deep-water sediments.

Phyllosilicates: A class of silicate minerals that form parallel sheets of silicate, where a central silicon atom is surrounded by four oxygen atoms at the corners of a tetrahedron. Three of the oxygen atoms of each tetrahedron are shared with other tetrahedrons. Examples are the clay minerals, kaolinite, and illite.

Total dissolved solids (TDS): Are a measure of the combined content of all inorganic and organic substances contained in water, as molecular, ionized, or micro-granular (colloidal sol) suspended form. The operational definition is that the solids must be small enough to survive filtration through a filter with 2 μm nominal size pores (or smaller). TDS are usually discussed only for freshwater systems.

Total suspended sediment (TSS): The portion of the sediment that is carried by a fluid flow, such as in a river or coastal current. It is maintained in suspension by turbulence in the flowing water and consists of particles generally of the fine sand, silt, and clay size.

References

Benedetti MF, Dia A, Riotte J et al (2003) Chemical weathering of basaltic lava flows undergoing extreme climatic conditions: the water geochemistry record. Chem Geol 201:1–17

Berner RA, Lassaga AC, Garrels RM (1983) The carbonate-silicate geochemical cycle and its effect on atmospheric carbon dioxide over the past 100 million years. Am J Sci 284:1183–1192

Biscaye PE (1965) Mineralogy and sedimentation of recent deep-sea clay in the Atlantic Ocean and adjacent seas and oceans. Geol Soc Am Bull 76:803–832

Boeglin JL, Probst JL (1998) Physical and chemical weathering rates and CO2 consumption in a tropical lateritic environment: the upper Niger basin. Chem Geol 148:137–156

Butman D, Raymond PA (2011) Significant efflux of carbon dioxide from streams and rivers in the United States. Nat Geosci. doi:10.1038/ngeo1294

Campbell IH, Reiners PW, Allen CM et al (2005) He-Pb dating of detrital zircons from the Ganges and Indus Rivers: implications for quantifying sediment recycling and provenance studies. Earth Planet Lett 237:402–432

Canfield DE (1997) The geochemistry of river particulates from continental USA: major elements. Geochim Cosmochim Acta 61(16):3349–3365

Depetris PJ, Pasquini AI (2007) The geochemistry of the Paraná River: an overview. In: Iriondo MH, Paggi JC, Parma MJ (eds) The middle Paraná River: limnology of a subtropical wetland. Springer, Berlin

Depetris PJ, Pasquini AI (2008) Riverine flow and lake level variability in southern South America. EOS Trans Am Geophys Union 89(28):254–255

Depetris PJ, Probst JL, Pasquini AI et al (2003) The geochemical characteristics of the Paraná River suspended sediment load: an initial assessment. Hydrol Proc 17:1267–1277

Depetris PJ, Gaiero DM, Probst JL, Hartmann J, Kempe S (2005) Biogeochemical output and typology of rivers draining Patagonia's Atlantic seaboard. J Coastal Res 21(4):835–844

Dupré B, Gaillardet J, Rousseau D et al (1996) Major and trace elements of river-borne material: the Congo Basin. Geochim Cosmochim Acta 60:1301–1321

Edmond JM, Huh Y (1997) Chemical weathering yields from basement and orogenic terrains in hot and cold climates. In: Ruddiman WF (ed) Tectonic uplift and climate change. Plenum Press, New York

Gaillardet J, Dupré B, Allègre CJ, et al. (1997) Chemical and physical denudation in the Amazon River basin. Chem Geol 142:141–173

Gaillardet J, Dupré B, Allègre CJ (1999) Geochemistry of large river suspended sediments: silicate weathering or recycling tracer? Geochim Cosmochim Acta 63(23/24):4037–4051

Gaillardet J, Viers J, Dupré B (2005) Trace elements in river waters. In: Drever JI (ed) Surface and ground water, weathering, and soils. Elsevier, Amsterdam

Garzanti E, Limonta M, Resentini A et al (2013) Sediment recycling at convergent plate margins (Indo-Burman Ranges and Andaman-Nicobar Ridge. Earth Sci Rev 123:113–132

Gibbs RJ (1970) Mechanism controlling world water chemistry. Science 170:1088–1090

Griffin JJ, Windom H, Goldgerg ED (1968) The distribution of clay minerals in the World Ocean. Deep-Sea Res 15:433–459

Holeman JN (1968) The sediment yield of major rivers of the world. Water Resour Res 4:737–747

Judson S, Ritter DF (1964) Rates of regional denudation in the US. J Geophys Res 69:3395–6401

Lecomte KL, Pasquini AI, Depetris PJ (2005) Mineral weathering in a semiarid mountain river: its assessment through PHREEQC inverse modeling. Aquatic Geochem 11:173–194

Lecomte KL, Milana JP, Formica SM et al (2008) Hydrochemical appraisal of ice- and rock-glacier meltwater in the hyperarid Agua Negra drainage basin, Andes of Argentina. Hydrol Proc 22:2180–2195

Lecomte KL, García MG, Formica SM et al (2009) Influence of geomorphological variables on mountainous stream water chemistry (Sierras Pampeanas de Córdoba, Argentina). Geomorphology 110:195–202

Ludwig W, Probst JL (1998) River sediment discharge to the oceans: present-day controls and global budgets. Am J Sci 298:265–295

McLennan SM, Hemming S, McDaniel DK et al (1993) Geochemical approaches to sedimentation, provenance, and tectonicts. Geol Soc Am Special Paper 284:21–40

Meybeck M (2005) Global occurrence of major elements in rivers. In: Drever JI (ed) Surface and ground water, weathering, and soils. Elsevier, Amsterdam

Milliman JD, Farnsworth KL (2011) River discharge to the coastal ocean. Cambridge University Press, Cambridge

Milliman JD, Meade RH (1983) World-wide delivery of river sediment to the oceans. J Geol 91:1–21

Milliman JD, Syvitski JPM (1992) Geomorphic/tectonic control of sediments discharge to the oceans: the importance of small mountainous rivers. J Geol 100:525–544

Nakkem M (1999) Wavelet analysis of rainfall-runoff variability isolating climatic and anthropogenic patterns. Environ Model Softw 14:283–295

Pasquini AI, Depetris PJ, Gaiero DM et al (2005) Material sources, chemical weathering, and physical denudation in the Chubut River basin (Patagonia, Argentina): Implications for Andean rivers. J Geol 113:451–469

Potter PE (1986) South America and a few grains of sand. Part I. Beach sands. J Geol 94:301–319

Pasquini AI, Depetris PJ (2007) Discharge trends and flow dynamics of South American rivers draining the southern Atlantic seaboard: An overview. J Hydrol 333(2–4):385–399

Potter PE, Hamblin WK (2006) Big rivers worldwide. Brigham Young University Geology Studies, Provo

Raymo NE, Ruddiman WF (1992) Tectonic forcing of late Cenozoic climate. Nature 359:117–122

Raymond PA, Cole JJ (2003) Increase in the export of alkalinity from North America's largest river. Science 301:88–91

Ruddiman WF (1997) Tectonic uplift and climate change. Plenum Press, New York

Syvitski JPM, Vörösmarty CJ, Kettner AJ et al (2005) Impact of humans on the flux of terrestrial sediment to the global coastal ocean. Science 308:376–380

Veizer J, Jansen SL (1979) Basement and sedimentary recycling and continental evolution. J Geol 87:341–370

Wohl E (2010) Mountain rivers revisited. American Geophysical Union, Washington DC

Chapter 7
Conclusion

Abstract Current data suggest that continents are denuded by rivers at a rate of $\sim$23 Gt yr^{-1}, most of which ($\sim$83 %) is accounted for by solid debris and the remaining fraction by dissolved solids. Aeolian dust and ice-rafted rock fragments contribute a less-known proportion. This material transfer does not necessarily reflect the intensity and rate of weathering occurring on the continents because a significant fraction is supplied by outcropping sedimentary rocks, thus intervening in the complex rock cycle. Continental denudation must, therefore, be considered in the context of the composite dynamics of the critical zone, which is significantly affected by anthropogenic interactions.

Keywords Weathering cycle · Denudation · Climate forcing · Anthropogenic forcing · Tectonic forcing · Critical zone · CO_2 sequestration · Erosion · Rivers · Exhumation

7.1 Final Remarks

Our main intention in the preparation of this volume for the Springer Briefs in Earth System Sciences was to present in a concise way, the complex chain of geological events that leads to the denudation of continents. The picture that we try to convey is that of a sequence that begins with mechanical weathering. Rock and chemical attack, appears often mediated by the subtle effect of biology. The whole process of continental wearing down is, in fact, an open system with internal feedbacks (e.g., incipient chemical weathering assists in the action of peripheral forces, such as frost weathering) and the powerful action of interacting external forcing, like climate and tectonics (Fig. 7.1). There is ample evidence that weathering affects climate, for example, sequestering CO_2, and climate, in turn, impacts on the rate at which weathering occurs (e.g., Ruddiman 1997).

Another aspect worthy of attention is the fact that, according to the most recent and comprehensive data set, continents appear to be denuded at a rate of nearly 23

P. J. Depetris et al., *Weathering and the Riverine Denudation of Continents*, SpringerBriefs in Earth System Sciences, DOI: 10.1007/978-94-007-7717-0_7,

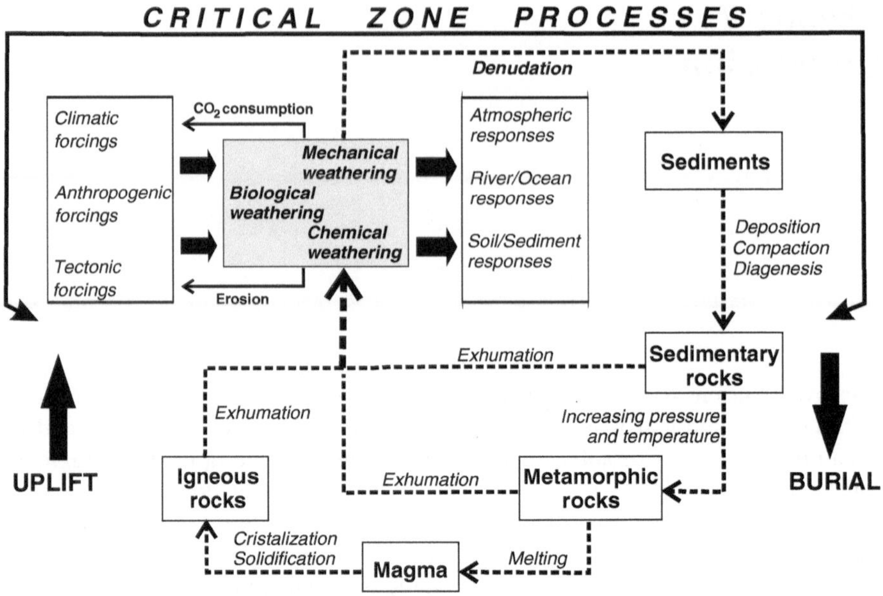

Fig. 7.1 Coupled chemical, physical, and biological weathering processes in the critical zone (embedded in the global geological cycle), which are affected by climate, anthropogenic and tectonic forcing over significantly different timescales. The output from the weathering engine is documented in the response of the atmosphere, hydrosphere, and geosphere

Gt yr^{-1} (Milliman and Farnsworth 2011), ~83 % of which appears to be accounted for by the sediment exported from the continental mass via rivers, and the remaining 17 % corresponds to the total solids that are delivered to the coastal oceans as dissolved matter. Are these figures in agreement with global weathering intensity and rate? Although they are sound approximations to continental wearing down, they do not reflect the significance of **current weathering** in a global perspective, as we will immediately see. Considering metamorphic rocks as either sedimentary or igneous depending on their origin, it is a common knowledge among earth scientists that the great bulk of the Earth's crust consists of igneous rocks (95 %) and only 5 % are sedimentary rocks, forming a relatively thin layer at or near the surface. However, the extent of sedimentary rocks cropping out at the Earth's surface is much larger than that of igneous rocks, so that 75 % of all rocks seen at the surface are sedimentary and only about 25 % are igneous. Therefore, as Gaillardet et al. (1999) have shown and other authors have suggested the sediment flux of a large proportion of the world's largest river systems is, in fact, a recycled material that has already passed once or several times through the Earth's exogenous cycle. Therefore, what the TSS nature and load of major large rivers show, as McLennan (1993) pointed out about 22 years ago, is the signature of the **weathering history** of any individual large river system.

Another aspect that has not been considered in this work is the significance of aeolian transport of dust to the oceans and the significance of ice-rafted debris. Clearly, both contribute to denude continents. However, knowledge on the former has increased significantly during the last decade and fluxes of continental aerosols transported to world oceans are known with increasing certainty for the present-day conditions, and for the recent geological past (e.g., Maher et al. 2010). Due to difficulties that are inherent to the process, the ice-rafted supply of sediment to the world's ocean is less known globally and restricted to specific areas (e.g., Jonkers et al. 2012).

As it is widely known, human actions play a mixed role; they increase erosion employing erroneous soil-use practices and, in contrast, they retain sediments behind dams, as seen in an earlier chapter. This aspect has been thoroughly dealt with by Milliman and Farnsworth (2011).

Concerning the linkage between chemical weathering and global denudation, there are two aspects that contribute to project a blurry image on this particular topic. Although there has been considerable progress on the knowledge concerning the global transfer of continental dissolved phases to the coastal oceans via ground water seepage there is still significant ground to cover. The knowledge is restricted to certain regions, although there is a growing certainty that such dissolved flux is, indeed, significant. The other aspect with still incomplete information derives from the global chemistry of rainfall and snowfall and the relative role played by recycled salts when computing global TDS continental fluxes.

Concluding this monograph, we hope that we have attained the goal that, in showing what is broadly known nowadays, we have also shown where scientists should dig deeper. In so doing, we tried to use updated references, modern concepts and, whenever possible, references to our own work, which was developed in the southern portion of South America.

References

Gaillardet J, Dupré B, Allègre CJ (1999) Geochemistry of large river suspended sediments: silicate weathering or recycling tracer? Geochim Cosmochim Acta 63(23/24):4037–4051

Jonkers L, Prins MA, Moros M et al (2012) Temporal offsets between surface temperature, ice-rafting and bottom flow speed proxies in the glacial (MIS 3) northern North Atlantic. Quatern Sci Rev 48:43–53

Maher BA, Prospero JM, Mackie D et al (2010) Global connections between aeolian dust, climate and ocean biogeochemistry at the present day and at the last glacial maximum. Earth Sci Rev 99(1–2):61–97

McLennan SM (1993) Weathering and global denudation. Jour Geol 101:295–303

Milliman JD, Farnsworth KL (2011) River discharge to the coastal ocean: a global synthesis. Cambridge University Press, Cambridge

Ruddiman WF (1997) Tectonic uplift and climate change. Plenum, New York

Index

P. J. Depetris et al., *Weathering and the Riverine Denudation of Continents*, SpringerBriefs in Earth System Sciences, DOI: 10.1007/978-94-007-7717-0,